CAMBRIDGE TRACTS IN MATHEMATICS

General Editors

137 Metric Diophantine Approximation on Manifolds

V. I. Bernik
Byelorussian Academy of Sciences

M. M. Dodson
University of York

Metric Diophantine Approximation on Manifolds

CAMBRIDGE
UNIVERSITY PRESS

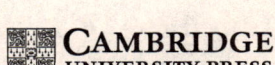

PUBLISHED BY THE PRESS SYNDICATE OF THE UNIVERSITY OF CAMBRIDGE
The Pitt Building, Trumpington Street, Cambridge, United Kingdom

CAMBRIDGE UNIVERSITY PRESS
The Edinburgh Building, Cambridge CB2 2RU, UK www.cup.cam.ac.uk
40 West 20th Street, New York, NY 10011–4211, USA www.cup.org
10 Stamford Road, Oakleigh, Melbourne 3166, Australia
Ruiz de Alarcón 13, 28014 Madrid, Spain

First published 1999

Printed in the United Kingdom at the University Press, Cambridge

Typeface Computer Modern 12/14pt *System* AmsTeX [UPH]

A catalogue record for this book is available from the British Library

ISBN 0 521 43275 8 hardback

To Haleh Afshar and Tatiana Bernik

Contents

Preface

This book is about metric Diophantine approximation on smooth manifolds embedded in Euclidean space. The aim is to develop a coherent body of theory on the lines of that which already exists for the classical theory, corresponding to the manifold being Euclidean space. Although the functional dependence of the coordinates presents serious technical difficulties, there is a surprising degree of interplay between the very different areas of number theory, differential geometry and measure theory.

A systematic theory began to emerge in the mid-1960's when V. G. Sprindžuk and W. M. Schmidt established that certain types of curve were extremal (an extremal set enjoys the property that, in a sense that can be made precise, Dirichlet's theorem on simultaneous Diophantine approximation cannot be improved for almost all points in the set; thus the real line is extremal). Sprindžuk conjectured that analytic manifolds satisfying a necessary nondegeneracy condition are extremal. Over the last 30 years, there has been considerable progress in verifying this conjecture for manifolds satisfying various arithmetic and geometric constraints, culminating in its recent proof by D. Y. Kleinbock and G. A. Margulis using ideas of flows on homogeneous spaces of lattices [139]. The greater part of this book is concerned with establishing the counterparts of Khintchine's theorem for manifolds and with the Hausdorff dimension of the associated exceptional sets. It relies very much on Sprindžuk's important monographs *Mahler's problem in metric number theory* [208] and *Metric theory of Diophantine approximations* [210]; indeed to some extent it can be regarded as a sequel. Our approach, like Sprindžuk's, is largely analytic and geometric and flows on lattices are not used nor, apart from the last chapter, is ergodic theory. These approaches, however, hold great promise even for the more delicate questions of Khintchine type results and Hausdorff dimension on manifolds.

Chapter 1 sets out the background required for metric Diophantine approximation on manifolds. Khintchine's theorem on simultaneous Diophantine approximation and its dual form (Groshev's theorem) are considered for manifolds in Chapter 2, which is devoted mainly to the long and demanding proof of a closely related conjecture of A. Baker concerning the rational normal curve $\{(t,\ldots,t^n)\colon t\in\mathbb{R}\}$. Chapter 3 begins with a relatively self-contained account of Hausdorff dimension and an introduction to its uses in Diophantine approximation. A fuller discussion is given in Chapters 4 and 5 which deal respectively with the technically different problems of obtaining upper and lower bounds. The range of techniques from the number theory arsenal which are called upon are an indication of the level of difficulty of some of the questions. The p-adic case is discussed fairly briefly in Chapter 6; the final chapter is devoted to various applications of metric Diophantine approximation.

Theorems, lemmas and so on are numbered consecutively in each chapter. Sections are denoted § $l.m$ and subsections by § $l.m.n$. The scope and complexity of the material has made notation something of a problem and to help the reader a list precedes Chapter 1. While not complete, the references are nevertheless intended to be reasonably comprehensive and include less well known papers from the former Soviet Union.

It is with sadness that we record that our friend and colleague Yuri Melnichuk would have been an author but for his tragic death during a visit to York in 1993. We are very grateful to many people and particularly to Haleh Afshar and Tatiana Bernik for their support and encouragement during this setback and throughout the book's lengthy gestation. Alan Baker has given us constant encouragement, Victor Beresnevich, Detta Dickinson, Sanju Velani, James Vickers and Chris Wood read parts of earlier drafts and made many suggestions and corrections; Peter Jackson read the proofs and removed numerous inconsistencies and typographical errors. They are not, however, responsible for any mistakes remaining.

The book was prepared on a Silicon Graphics Personal Iris workstation using LaTeX and GNU Emacs installed by Michael Beaty who with Simon Eveson sorted out our TeX problems with skill and good humour. Roger Astley of the Cambridge University Press has been patient and understanding beyond the call of duty. The collaboration essential to this book would not have been possible without the support that the Royal Society and the Soros Foundation provided for exchanges between the Belorussian Academy of Sciences at Minsk, the Lvov Polytechnic Institute and the University of York. The help which we have had has been invaluable and has ensured that this book will be published this side of the millennium.

V. I. Bernik
Institute of Mathematics
Academy of Sciences
Minsk, Belarus

M. M. Dodson
Department of Mathematics
University of York
York, UK

Notation

CHAPTER 1

Diophantine approximation and manifolds

1.1. Introduction

Diophantine approximation is a more quantitative and general study of the density of the rationals $\mathbb{Q}$ in the reals $\mathbb{R}$ while a smooth manifold is locally diffeomorphic to Euclidean space. In this chapter, those parts of Diophantine approximation and differential geometry needed are set out. The former is concerned mainly with the inequality

$$\left|\xi - \frac{p}{q}\right| < \varepsilon,$$

where ξ is a real number and ε is a small positive number depending on the rational p/q, and its higher dimensional versions. In the metric theory, solution sets of Diophantine inequalities are considered in terms of Lebesgue measure (a knowledge of this is assumed). Because an exceptional set for which a result is false can be of measure zero, this can lead to theorems, such as Khintchine's theorem below, having a strikingly simple yet general character. Moreover the exceptional sets can in turn be analysed in terms of Hausdorff dimension. The analysis becomes much more difficult when considering points on a manifold in Euclidean space, as the coordinates are functionally related and so dependent.

Much of the material in this book is a further development of Sprindžuk's monograph [210] which, starting with a thorough discussion of Khintchine's theorem and its generalisations, goes on to a systematic account of the emergent theory of metric Diophantine approximation on manifolds. J. W. S. Cassels' tract [59] contains a concise but comprehensive introduction to Diophantine approximation and G. Harman's recent book *Metric number theory* [115] has a wider scope which includes brief accounts of Diophantine approximation on manifolds (Chapter 9) and Hausdorff dimension (Chapter 10).

1.2. Diophantine approximation in one dimension

Dirichlet's theorem is fundamental to the theory of Diophantine approximation. The one dimensional form of the theorem states that for each real number ξ and any positive integer N, there exists a rational p/q with positive denominator $q \leqslant N$, such that

$$\left|\xi - \frac{p}{q}\right| < \frac{1}{qN}.$$

Since $q \leqslant N$, it immediately follows that

$$\left|\xi - \frac{p}{q}\right| < \frac{1}{q^2}. \tag{1.1}$$

If p and q are not restricted to being coprime, then there are infinitely many solutions, otherwise there are only finitely many solutions when $\xi \in \mathbb{Q}$ (for further details see [59], [114], [201]).

It is convenient to introduce some notation. As usual, $\mathbb{N}$ will denote the positive integers $1, 2, \ldots$, $\mathbb{Z}$ will denote the integers, the integer part of the real number ξ is the greatest integer at most ξ and will be denoted by $[\xi]$. The fractional part $\xi - [\xi]$ of ξ is non-negative and is written $\{\xi\}$. A standard and simplifying notation which places the denominator q in the foreground is to write

$$\|\xi\| = \min\{|\xi - r| \colon r \in \mathbb{Z}\} = \min\{\{\xi\}, \{1 - \xi\}\},$$

so that (1.1) becomes $\|q\xi\| < 1/q$. Note that $\|\xi + \xi'\| \leqslant \|\xi\| + \|\xi'\|$ and that $\|r\xi\| \leqslant |r|\|\xi\|$ when $r \in \mathbb{Z}$. The *symmetrised* fractional part of ξ defined by

$$\langle \xi \rangle = \begin{cases} \{\xi\} & \text{when } 0 \leqslant \{\xi\} \leqslant 1/2, \\ \{\xi\} - 1 & \text{otherwise,} \end{cases}$$

lies in $(-1/2, 1/2]$ and satisfies $\|\xi\| = |\langle \xi \rangle|$. Given positive real numbers a, b, the Vinogradov notation

$$a \ll b \text{ or } b \gg a$$

is used for $a = O(b)$, *i.e.*, when $a \leqslant Kb$ for some positive constant K. If $a \ll b$ and $a \gg b$, a and b are said to be *comparable*, denoted by $a \asymp b$.

1.2.1. Approximation functions. More generally, let $\psi \colon \mathbb{N} \to \mathbb{R}^+$ be a positive function ($\mathbb{R}^+ = \{x \in \mathbb{R} \colon x > 0\}$) where $\psi(q) \to 0$ as $q \to \infty$. Let $X \subseteq \mathbb{R}$. We will write $\mathscr{K}(X; \psi)$ for the set of $\xi \in X$ such that the more general inequality

$$\|q\xi\| < \psi(q) \tag{1.2}$$

holds for infinitely many positive integers q, *i.e.*,

$$\mathscr{K}(X; \psi) = \{\xi \in X \colon \|q\xi\| < \psi(q) \text{ for infinitely many } q \in \mathbb{N}\}$$

first studied by A. I. Khintchine [134]. Points in $\mathscr{K}(X; \psi)$ will be called *ψ-approximable*. When the set X is clear from the context, we will usually omit reference to it and write simply $\mathscr{K}(\psi)$. The function ψ will be called an *approximation function* and will often be taken to be monotonically decreasing (we will usually omit the term monotonically) as well. Note that $\psi(q) \leqslant 1/2$ when q is sufficiently large. We will make much use of the observation that the set $\mathscr{K}(X;\psi)$ and its generalisations are 'lim-sup' sets as

$$\begin{aligned}\mathscr{K}(X;\psi) &= \{\xi \in X : \xi \in B_{\psi(q)}(q) \text{ for infinitely many } q \in \mathbb{N}\} \\ &= \bigcap_{N=1}^{\infty} \bigcup_{q=N}^{\infty} B_{\psi(q)}(q) = \limsup_{N\to\infty} B_{\psi(N)}(N),\end{aligned}$$

where $B_\delta(q) = \{\xi \in X : \|q\xi\| < \delta\}$. In particular they are Borel sets [100].

The expression $\|\xi\|$ is invariant under translation by integers so that given any integer r, $\xi + r$ satisfies (1.2) if and only if ξ does. Thus $\mathscr{K}([0,1]+r;\psi) = \mathscr{K}([0,1];\psi) + r$ and

$$\mathscr{K}(\mathbb{R};\psi) = \bigcup_{r\in\mathbb{Z}} (\mathscr{K}([0,1];\psi) + r)\,.$$

When considering the measure of the set of ψ-approximable real numbers, there is of course no loss in generality in considering points in any (proper) interval.

In the important special case when $\psi(q) = q^{-v}$, we write $\mathscr{K}_v(X)$ for $\mathscr{K}(X;\psi)$; thus

$$\mathscr{K}_v(X) = \{\xi \in X : \|q\xi\| < q^{-v} \text{ for infinitely many } q \in \mathbb{N}\}.$$

Dirichlet's theorem implies that $\mathscr{K}_1(\mathbb{R}) = \mathbb{R}$. Points in $\mathscr{K}_v(\mathbb{R})$ are called *v-approximable*; there should be no confusion with ψ-approximable points. If a point lies in $\mathscr{K}_v(\mathbb{R})$ for some $v > 1$, it is called *very well approximable* [201]. Thus the set of very well approximable points is the union of v-approximable points for $v > 1$. The related set $\mathcal{K}_v(X)$ is defined as follows. For each $\xi \in \mathbb{R}$, let

$$\omega(\xi) = \sup\{w \in \mathbb{R} : \xi \in \mathscr{K}_w(\mathbb{R})\}$$

($\omega(\xi) \geqslant 1$ by Dirichlet's theorem). For any set $X \subseteq \mathbb{R}$ and $v \in \mathbb{R}$, write

$$\mathcal{K}_v(X) = \{\xi \in X : \omega(\xi) \geqslant v\}. \tag{1.3}$$

It is readily verified that $\mathscr{K}_v(X) \subseteq \mathcal{K}_v(X) \subseteq \mathscr{K}_{v-\varepsilon}(X)$ for any $\varepsilon > 0$. The nature of the approximation function in $\mathcal{K}_v(X)$ enables one to analyse $\mathcal{K}'_v(X)$, the set of ξ with $\omega(\xi) = v$ (see §3.5.6).

1.2.2. Badly approximable numbers. A number ξ is *badly approximable* if there exists a positive constant $K = K(\xi)$ such that

$$\|q\xi\| \geqslant K/q$$

for all $q \in \mathbb{N}$, *i.e.*, if $\|q\xi\| \gg 1/q$ (but see the Notes and [210, p. 67]). The set of badly approximable numbers will be denoted by $\mathfrak{B}$. By Hurwitz' theorem, for each $\xi \in \mathbb{R}$ the inequality

$$\|q\xi\| < \frac{1}{\sqrt{5}q}$$

holds for infinitely many positive integers q. However, the constant $1/\sqrt{5}$ cannot be reduced for numbers ξ equivalent to the golden ratio $(\sqrt{5}-1)/2$. These numbers are thus badly approximable (see [59], [114] for further details). Badly approximable numbers are important in applications, particularly in stability questions for certain dynamical systems (see Chapter 7), but they are amply covered in [201] and so will not be discussed in any detail.

1.2.3. Khintchine's theorem. The behaviour of the sum $\sum_{q=1}^{\infty}\psi(q)$ gives an almost complete answer to the solubility of the inequality (1.2). First we need some terminology. A set of Lebesgue measure 0 will usually be called *null*; the complement of a null set is of *full measure* and will usually be called *full*. As usual we will say that *almost no* points belong to a set if it is null while if a set is full we say that it contains *almost all* points. The Lebesgue measure of a set X will be denoted by $|X|$.

THEOREM (KHINTCHINE). *Let $\psi : \mathbb{N} \to \mathbb{R}^+$ be a function. If the sum $\sum_{q=1}^{\infty}\psi(q)$ converges, then $\mathscr{K}(\mathbb{R};\psi)$ is null, while if the sum diverges and ψ is decreasing, $\mathscr{K}(\mathbb{R};\psi)$ is full.*

Proofs can be found in [59, Chapter VII], [115, Chapter 2] and [210]. In the case of convergence, the result essentially follows from the Borel-Cantelli lemma. Since Cantelli pointed out that the total independence of the events was not needed for convergence [64, p. 507], we will refer for brevity to the convergence part of the lemma as Cantelli's lemma. This will now be stated and proved as it will be used repeatedly throughout.

LEMMA 1.1 (CANTELLI). *Let (Ω,μ) be a measure space with $\mu(\Omega)$ finite and let A_j, $j \in \mathbb{N}$, be a family of measurable sets. Let*

$$A_\infty = \{\omega \in \Omega \colon \omega \in A_j \text{ for infinitely many } j \in \mathbb{N}\}$$

and suppose the sum $\sum_{j=1}^{\infty}\mu(A_j) < \infty$. Then $\mu(A_\infty) = 0$.

PROOF. It is readily verified that A_∞ can be written in 'lim-sup' form as

$$A_\infty = \bigcap_{N=1}^{\infty}\bigcup_{j=N}^{\infty} A_j.$$

It follows that for each $N = 1, 2, \ldots$, the family $\{A_j \colon j \geqslant N\}$ is a cover for the set A_∞, so that $A_\infty \subseteq \bigcup_{j=N}^{\infty} A_j$, whence

$$\mu(A_\infty) \leqslant \sum_{j=N}^{\infty}\mu(A_j).$$

But the sum $\sum_{j=1}^{\infty}\mu(A_j)$ converges whence the tail $\sum_{j=N}^{\infty}\mu(A_j)$ of the series can be made arbitrarily small and so $\mu(A_\infty) = 0$. □

For each $N = 1, 2, \ldots$, the family $\{A_j : j \geqslant N\}$ will be called the *natural* cover for A_∞.

To deduce Khintchine's theorem in the case of convergence, we recall that without loss of generality we can restrict ourselves to the set $[0,1]$. Take $\Omega = [0,1]$ and μ to be Lebesgue measure. Any point in the set $\mathscr{K}([0,1]\,;\psi)$ lies in infinitely many sets $B_{\psi(q)}(q)$, where

$$B_\delta(q) = \{\xi \in [0,1] : \|q\xi\| < \delta\} = \bigcup_{p=0}^{q} \left(\frac{p}{q} - \frac{\delta}{q}, \frac{p}{q} + \frac{\delta}{q}\right) \cap [0,1]$$

and the family $\{B_{\psi(q)}(q) : q \in \mathbb{N}\}$ is a natural cover for $\mathscr{K}([0,1]\,;\psi)$. Each $B_\delta(q)$ is a union of $q+1$ open intervals. By adding up the lengths of the intervals it can be seen that $|B_\delta(q)| \leqslant 2\delta$ (with equality when $\delta \leqslant 1/2$). Hence the sum $\sum_q |B_{\psi(q)}(q)| \leqslant 2\sum_q \psi(q)$ converges and the result follows from Cantelli's lemma. The case of divergence is harder and a monotonicity condition on the approximation function ψ is required (more details are in [59], [115], [210]); a brief discussion of the more general theorem is in §1.3.4 below.

The theorem corresponds to our intuition since if the approximation function ψ is large then there is a better chance of the inequality being satisfied. Since the sum $\sum_{r=1}^{\infty} r^{-v}$ converges for $v > 1$ and diverges otherwise, Khintchine's theorem implies that the sets $\mathscr{K}_v([0,1])$ and $\mathcal{K}_v([0,1])$ are null or full according as $v > 1$ or $v \leqslant 1$ respectively (in the latter case they are both the real line by Dirichlet's theorem). Less obviously, the theorem shows the Lebesgue measure of the set of $\xi \in [0,1]$ such that (1.2) has infinitely many solutions is 1 when $\psi(q) = 1/(q \log q)$ and 0 when $\psi(q) = 1/(q(\log q)^{1+\varepsilon})$ for any positive ε. This 'zero-one' property is a feature of the metric theory and reflects the links with probability and ergodic theory.

The theorem also implies that the set $\mathfrak{B}$ of badly approximable numbers is null. For given any $K > 0$, the sum $\sum_q (K/q)$ diverges and so by Khintchine's theorem the set of real numbers ξ satisfying $\|q\xi\| < K/q$ for infinitely many $q \in \mathbb{N}$ is full. Thus the complementary set $V(K)$ of ξ such that $\|q\xi\| \geqslant K/q$ for all but finitely many q is null and evidently increases as K decreases. From its definition,

$$\mathfrak{B} \subset \bigcup_{K>0} V(K) = \bigcup_{N=1}^{\infty} V(1/N),$$

a countable union of null sets, whence $\mathfrak{B}$ is null.

1.3. Approximation in higher dimensions

The inequality (1.2) can be generalised to higher dimensions. To describe these generalisations concisely, we set down some notation. Throughout, m, n and N will be positive integers, k, p, q will be integers and q will usually be taken to be positive. *Integer* vectors in Euclidean space will always be written with a bold font, thus $\mathbf{q} \in \mathbb{Z}^n$. The *height* or supremum norm $|\xi|_\infty$ of the vector $\xi \in \mathbb{R}^n$ will be denoted by $|\xi|$, so that

$$|\xi| = \max\{|\xi_1|, \ldots, |\xi_n|\}.$$

To some extent in number theory the height replaces the usual Euclidean norm which will be written $|\xi|_2$. The inner or scalar product of two vectors ξ and ζ will be written $\xi \cdot \zeta$. The symmetrised fractional part of a vector $\xi = (\xi_1, \ldots, \xi_n)$ in $\mathbb{R}^n$ is defined by

$$\langle \xi \rangle = (\langle \xi_1 \rangle, \ldots, \langle \xi_n \rangle) \in (-1/2, 1/2]^n \tag{1.4}$$

and should not be confused with the inner product. Note that there is a unique $\mathbf{k}_\xi \in \mathbb{Z}^n$ such that $\xi - \mathbf{k}_\xi = \langle \xi \rangle$.

The system

$$\xi_1 a_{1j} + \cdots + \xi_m a_{mj}, \; 1 \leqslant j \leqslant n,$$

of n real linear forms in m variables $\xi_1, \ldots, \xi_m$ will be written more concisely in matrix form as ξA, where $\xi \in \mathbb{R}^m$ and where by juxtaposing the m rows, the $m \times n$ real matrix $A = (a_{ij})$ is regarded as a point in $\mathbb{R}^{mn}$, *i.e.*, the space of $m \times n$ real matrices is identified with $\mathbb{R}^{mn}$. The inequality (1.2) can be generalised to the system of inequalities

$$|\langle q_1 a_{1j} + \cdots + q_m a_{mj} \rangle| = \|q_1 a_{1j} + \cdots + q_m a_{mj}\| < \psi(|\mathbf{q}|), \quad 1 \leqslant j \leqslant n.$$

Using the notation above, this system can be expressed as

$$|\langle \mathbf{q} A \rangle| < \psi(|\mathbf{q}|). \tag{1.5}$$

Matrices satisfying this inequality for infinitely many integer vectors $\mathbf{q}$ are called *ψ-approximable* [127]. The extension of Dirichlet's theorem to higher dimensions as a system of simultaneous inequalities involving linear forms is now stated.

THEOREM 1.2. *Let $A = (a_{ij})$ be an $m \times n$ real matrix. For each real $N > 1$, there exists an integer vector $\mathbf{q} \in \mathbb{Z}^m$ with $1 \leqslant |\mathbf{q}| < N$ such that*

$$|\langle \mathbf{q} A \rangle| < N^{-m/n}.$$

Proofs using Minkowski's linear forms theorem are in [59, p. 13, Theorem VI]; similar results using box arguments are in [114], [201].

1.3.1. Khintchine's transference principle. Simultaneous and dual Diophantine approximation are related by a 'transference' principle in which a solution in one form is related to a solution in the other form (or more accurately, the form associated with the transpose of the matrix of coefficients). This principle enables information about linear form inequalities and simultaneous Diophantine approximation to be interchanged to a certain extent. In particular it links simultaneous Diophantine approximation on the rational normal curve

$$\mathscr{V} = \mathscr{V}^{(n)} = \{(t, t^2, \ldots, t^n) \colon t \in I\} \tag{1.6}$$

over the interval I with the dual form and so with the distribution of small values of integral polynomials $P(t) = a_0 + a_1 t + \cdots + a_n t^n$, $a_0, \ldots, a_n \in \mathbb{Z}$ for $t \in I$ (see §1.4.4).

THEOREM (KHINTCHINE'S TRANSFERENCE PRINCIPLE). *Suppose the coordinates of $\xi = (\xi_1, \ldots, \xi_n) \in \mathbb{R}^n$ are irrational and let $\omega(\xi) \geqslant 0$, $\omega'(\xi) \geqslant 0$ be the respective least upper bounds of the real numbers w, w' for which the inequalities*

$$\|q_1\xi_1 + \cdots + q_n\xi_n\| \leqslant (\max |q_i|)^{-n-w},$$
$$\max_{1 \leqslant j \leqslant n} \|q\xi_j\| = |\langle q\xi \rangle| \leqslant q^{-(1+w')/n}$$

have infinitely many integer solutions. Then

$$\frac{\omega(\xi)}{n^2 + (n-1)\omega(\xi)} \leqslant \omega'(\xi) \leqslant \omega(\xi)$$

with the obvious interpretation if $\omega(\xi)$ or $\omega'(\xi)$ is infinite.

For a proof, see [59, Chapter 5, Theorem IV]. This transference principle implies that given any $\varepsilon > 0$, if $\xi = (\xi_1, \ldots, \xi_n)$ satisfies $|\langle \mathbf{q} \cdot \xi \rangle| < |\mathbf{q}|^{-n-\varepsilon}$ for infinitely many $\mathbf{q} \in \mathbb{Z}^n$, then for some ε' comparable to ε, $|\langle q\xi \rangle| < q^{-(1+\varepsilon')/n}$ for infinitely many $q \in \mathbb{Z}$, and vice versa. Note that the smaller the modulus of ω, ω', the more complete is the interchange of information.

1.3.2. Two forms of Diophantine approximation. We will be concerned mainly with the two special cases of the general inequality (1.5), namely when A is a $1 \times n$ real matrix or a $n \times 1$ real matrix (in both cases we regard A as a vector in $\mathbb{R}^n$). A natural question is whether in higher dimensions, subsets such as curves or surfaces, enjoy arithmetic approximation properties corresponding to those for real numbers and $\mathbb{R}^n$.

Let $\xi \in \mathbb{R}^n$ and suppose $\psi(q) \leqslant 1/2$ for all sufficiently large $q \in \mathbb{N}$. First, a point ξ satisfying the system of simultaneous inequalities

$$|\langle q\xi \rangle| = \max\{\|q\xi_1\|, \ldots, \|q\xi_n\|\} < \psi(q) \tag{1.7}$$

lies within (in the sup metric) $\psi(q)/q$ of the point $\mathbf{p}/q$, *i.e.*,

$$\xi \in \{x \in \mathbb{R}^n \colon |x - \mathbf{p}/q| < \psi(q)/q\},$$

where $\mathbf{p} \in q\,\xi + (-\psi(q), \psi(q))^n$ and is unique when $\psi(q) \leqslant 1/2$. Given a set X in $\mathbb{R}^n$ (later X will be taken to be a manifold), the set of $x \in X$ satisfying (1.7) for infinitely many positive integers q will be denoted by

$$\mathscr{S}(X;\psi) = \{\xi \in X \colon |\langle q\xi \rangle| < \psi(q) \text{ for infinitely many } q \in \mathbb{N}\}. \tag{1.8}$$

Points in $\mathscr{S}(X;\psi)$ will be called *simultaneously ψ-approximable.*

Secondly we consider the 'transposed' or 'dual' inequality

$$|\langle \mathbf{q} \cdot \xi \rangle| = \|\mathbf{q} \cdot \xi\| = \|q_1\xi_1 + \cdots + q_n\xi_n\| < \psi(|\mathbf{q}|), \tag{1.9}$$

involving the linear form $\|\mathbf{q}\cdot\xi\|$. Here the point ξ is within (in the Euclidean metric) $\psi(|\mathbf{q}|)/|\mathbf{q}|_2$ of the hyperplane

$$\{x \in \mathbb{R}^n : \mathbf{q}\cdot x = p\},$$

where $p \in \mathbf{q}\cdot\xi + (-\psi(|\mathbf{q}|), \psi(|\mathbf{q}|))$ is unique when $\psi(|\mathbf{q}|) \leqslant 1/2$. The set of $\xi \in X$ satisfying (1.9) for infinitely many integer vectors $\mathbf{q}$ will be denoted by $\mathscr{L}(X;\psi)$,

$$\mathscr{L}(X;\psi) = \{\xi \in X : |\langle \mathbf{q}\cdot\xi\rangle| < \psi(|\mathbf{q}|) \text{ for infinitely many } \mathbf{q} \in \mathbb{Z}^n\}. \quad (1.10)$$

Simultaneous Diophantine approximation has a historical priority and it is convenient to refer to the last inequality as the *dual* inequality. Thus points in $\mathscr{L}(X;\psi)$ will be called *dually ψ-approximable.* When there is no risk of confusion, we will refer just to ψ-approximable points. Note that $\mathscr{S}(X;\psi)$ and $\mathscr{L}(X;\psi)$ are lim-sup sets and that when $X \subseteq \mathbb{R}$, $\mathscr{S}(X;\psi) = \mathscr{L}(X;\psi) = \mathscr{K}(X;\psi)$.

In the important case when $\psi(r) = r^{-v}$, we write $\mathscr{S}_v(X)$ for $\mathscr{S}(X;\psi)$ for $\mathscr{L}(X;\psi)$. The sets $\mathscr{S}_v(X)$ and $\mathscr{L}_v(X)$ decrease as v increases and when $X = \mathbb{R}^n$ are null for $v > 1/n$ and $v > n$ respectively [59, Chapter 1]. A point in $\mathscr{S}_v(X)$ will be called *simultaneously v-approximable* and a point in $\mathscr{L}_v(X)$ will be called *dually v-approximable.* By Khintchine's transference principle, if a point in $\mathbb{R}^n$ is simultaneously v-approximable for $v > 1/n$, then it is dually v'-approximable for some $v' > n$, and vice versa. When v is close to $1/n$, v' is close to n but, as v gets larger, v' gets much further from n and indeed

$$\bigcup_{v>1/n} \mathscr{S}_v(X) = \bigcup_{v>n} \mathscr{L}_v(X). \quad (1.11)$$

Points in this set are called *very well approximable* and the set of such points in X is a countable union of the sets $\mathscr{S}_{1/n+1/r}(X)$ or $\mathscr{L}_{n+1/r}(X)$, $r = 1, 2, \ldots$.

Dirichlet's theorem in higher dimensions specialises to simultaneous Diophantine approximation and to the 'dual' linear form.

COROLLARY 1.3. *For each $\xi = (\xi_1, \ldots, \xi_n) \in \mathbb{R}^n$ there exist*
(a) an integer q with $1 \leqslant q \leqslant N$ and a vector $\mathbf{p} \in \mathbb{Z}^n$ such that

$$|q\xi - \mathbf{p}| < N^{-1/n} \quad (1.12)$$

and there are infinitely many positive integers q such that

$$|\langle q\xi\rangle| < q^{-1/n},$$

(b) an integer vector $\mathbf{q} \in \mathbb{Z}^n$ with $1 \leqslant |\mathbf{q}| \leqslant N$ and an integer p such that

$$|\mathbf{q}\cdot\xi - p| < N^{-n}$$

and there are infinitely many $\mathbf{q} \in \mathbb{Z}^n$ such that

$$\|\mathbf{q}\cdot\xi\| < |\mathbf{q}|^{-n}.$$

Hence any point in $\mathbb{R}^n$ is simultaneously $(1/n)$-approximable and dually n-approximable, so that for any $X \subseteq \mathbb{R}^n$, $\mathscr{S}_{1/n}(X) = \mathscr{L}_n(X) = X$.

The notion of a badly approximable number can also be extended to Euclidean space. The point $\xi \in \mathbb{R}^n$ is *badly approximable* if there exists a positive number K such that

$$|\langle q\xi \rangle| \geqslant K q^{-1/n}$$

for all $q \in \mathbb{N}$; or equivalently by Khintchine's transference principle if there exists a $K' > 0$ such that

$$\|\mathbf{q} \cdot \xi\| \geqslant K' |\mathbf{q}|^{-n}$$

for all non-zero $\mathbf{q} \in \mathbb{Z}^n$. The set of badly approximable points in Euclidean space is null [59, Chapter 1].

The Diophantine approximation considered so far has been homogeneous. The rather different inhomogeneous approximation where for example given $\alpha \in \mathbb{R}$, one considers the inequality $|q\xi - p - \alpha| < \psi(q)$, will not be covered but some further details and references are in the Notes at the end of this chapter and of Chapter 3.

1.3.3. Order and exponents of approximation. A real number ξ which for some $K > 0$ satisfies the inequality

$$\left| \xi - \frac{p}{q} \right| < \frac{K}{q^n}$$

for infinitely many rationals p/q is called *(rationally) approximable* to order n [114, §11.2]. Thus if ξ is rationally approximable to order $r + 1$, the inequality

$$\|q\xi\| < K q^{-r}$$

holds for infinitely many positive integers q. By Dirichlet's theorem every real number can be approximated to order 2. This definition extends naturally to any real exponent and to simultaneous Diophantine approximation and the dual form.

Information about points with exponent of approximation v can be obtained when the approximation function $\psi(q)$ is a power of the form q^{-v}. Given a subset X of $\mathbb{R}^n$ and a point $\xi \in X$, let

$$\omega_{\mathscr{S}}(\xi) = \sup\{w \in \mathbb{R} \colon \xi \in \mathscr{S}_w(X)\} \tag{1.13}$$

$$\omega_{\mathscr{L}}(\xi) = \sup\{w \in \mathbb{R} \colon \xi \in \mathscr{L}_w(X)\}. \tag{1.14}$$

Note that if $X = \mathbb{R}^n$, then by Dirichlet's theorem, $\omega_{\mathscr{S}}(\xi) \geqslant 1/n$ and $\omega_{\mathscr{L}}(\xi) \geqslant n$. When $X \subseteq \mathbb{R}^n$ and each real v, the set

$$\mathcal{S}_v(X) = \{\xi \in X \colon \omega_{\mathscr{S}}(\xi) \geqslant v\}$$

$$\mathcal{L}_v(X) = \{\xi \in X \colon \omega_{\mathscr{L}}(\xi) \geqslant v\}$$

The sets $\mathcal{S}'_v(X)$ where $\omega_{\mathscr{S}}(\xi) = v$ and $\omega_{\mathscr{L}}(\xi) = v$ respectively in the definitions above, are higher dimensional versions of $\mathcal{K}'_v$.

Also $\mathcal{S}_{1/n}(X) = \mathcal{L}_n(X) = X$ and it can be verified readily that for any $\varepsilon > 0$,

$$\mathscr{S}_v(X) \subseteq \mathcal{S}_v(X) \subseteq \mathscr{S}_{v-\varepsilon}(X) \text{ and } \mathscr{L}_v(X) \subseteq \mathcal{L}_v(X) \subseteq \mathscr{L}_{v-\varepsilon}(X). \tag{1.15}$$

When $n = 1$, $\mathcal{S}_v(\mathbb{R}) = \mathcal{L}_v(\mathbb{R}) = \mathcal{K}_v(\mathbb{R})$.

In order to have the same definition for both types of approximation, we say that a point $\xi \in \mathbb{R}^n$ is *simultaneously approximable to exponent* v if there exists a constant K such that

$$|\langle q\,\xi\rangle| < Kq^{-v}$$

for infinitely many positive integers q; and is *dually approximable to exponent* v if there exists a K' such that

$$\|\mathbf{q}\cdot\xi\| < K'|\mathbf{q}|^{-v}$$

for infinitely many $\mathbf{q} \in \mathbb{Z}^n$.

It follows from Khintchine's Transference Principle that for any $X \subseteq \mathbb{R}^n$,

$$|\mathcal{L}_{nv/(1-(n-1)v)}(X)| \leqslant |\mathcal{S}_v(X)| < |\mathcal{L}_{n(1+v)-1}(X)|.$$

The order of approximation can be made more precise. The set

$$\mathcal{S}'_v(X) = \{\xi \in \mathcal{S}(X)\colon \omega_{\mathscr{S}}(\xi) = v\}, \tag{1.16}$$

where $\omega_{\mathscr{S}}(x)$ is given by (1.13), is the set of points in X which can be approximated simultaneously with exact exponent v (but order $v+1$). Similarly, for each $v \in \mathbb{R}$, let

$$\mathcal{L}'_v(X) = \{\xi \in \mathcal{L}(X)\colon \omega_{\mathscr{L}}(\xi) = v\}, \tag{1.17}$$

where $\omega_{\mathscr{L}}(x)$ is given by (1.14). Then $\mathcal{L}'_v(X)$ is the set of points in X which can be approximated dually with exact exponent v. The relationship between $\mathscr{S}_v$, $\mathcal{S}_v$ and $\mathcal{S}'_v$ and between $\mathscr{L}_v$, $\mathcal{L}_v$ and $\mathcal{L}'_v$ will be treated further in Chapter 3. The nature of the sets $\mathcal{S}'_v(X)$ and $\mathcal{L}'_v(X)$ allows their Hausdorff dimension to be determined exactly (this is discussed in §3.5.6).

1.3.4. The Khintchine-Groshev theorem. A very general form of Khintchine's theorem was obtained by A. V. Groshev (see [210, Chapter 1,§5]). As in the one dimensional case, this result gives precise information about the Lebesgue measure of the set $W(X;\psi)$ of ψ-approximable points in the set X when X is Euclidean space or a hypercube (when X is clear, we write simply $W(\psi)$). Further details are given in [83], [210].

THEOREM (GROSHEV). *Let ψ be a function from $\mathbb{N} \to \mathbb{R}^+$. Suppose the sum*

$$\sum_{r=1}^{\infty} r^{m-1}\psi(r)^n \tag{1.18}$$

converges. Then almost no points $A \in \mathbb{R}^{mn}$ satisfy

$$|\langle \mathbf{q}A\rangle| < \psi(|\mathbf{q}|) \tag{1.19}$$

for infinitely many $\mathbf{q} \in \mathbb{Z}^m$. *On the other hand suppose that the sum* (1.18) *diverges and that* $r^{m-1}\psi^n(r)$ *decreases for* r *large when* $m = 1, 2$. *Then almost all matrices* $A \in \mathbb{R}^{mn}$ *satisfy the inequality* (1.19) *for infinitely many* $\mathbf{q} \in \mathbb{Z}^m$.

Since $\langle \mathbf{q}A \rangle$ is invariant under translation by integer matrices, there is no loss in generality in confining attention to I^{mn}, where I is an interval of unit length. This has the advantage of making a probabilistic interpretation of the theorem more natural. When the sum (1.18) converges, the proof of the theorem is straightforward and, as in the one dimensional case, follows from Cantelli's lemma. The divergent case is much harder and when $m = 1$, the proof involves second moments [59, Chapter VII] or pairwise quasi-independence and ergodic theory [210, Chapter 1], [214]. Analytic proofs for systems of linear forms are in [83], [210, §2]; the monotonicity conditions can be relaxed when $m \geqslant 3$.

When the sum diverges, there is an asymptotic formula with leading term comparable to the partial sum up to Q of (1.18) for the number of $\mathbf{q} \in \mathbb{Z}^m$ of height at most Q such that (1.19) holds [196], [201, Chapter III]. This topic is not covered in this book but more details can be found in [210, Chapter 2, §12] and [115]. In the case of convergence, the set $W(\psi)$ is null; its Hausdorff dimension is determined in Chapters 4 and 5 below.

Simultaneous and dual Diophantine approximation are special cases of the theorem.

COROLLARY 1.4. *Let* $\psi \colon \mathbb{N} \to \mathbb{R}^+$ *be an approximation function. Then*

(a) the set $\mathscr{S}(\mathbb{R}^n; \psi)$ *is null or full according as* $\sum_{q=1}^{\infty} \psi(q)^n$ *converges or diverges and* ψ *is decreasing;*

(b) the set $\mathscr{L}(\mathbb{R}^n; \psi)$ *is null or full according as* $\sum_{q=1}^{\infty} q^{n-1}\psi(q)$ *converges or diverges and* $q^{n-1}\psi(q)$ *is decreasing when* $n = 1, 2$.

Thus apart from some minor restrictions on the approximation function ψ, the sets $\mathscr{S}(X; \psi)$ and $\mathscr{L}(X; \psi)$ are well understood in terms of Lebesgue measure when the coordinates (or 'variables') $\xi_1, \dots, \xi_n$ of the points $\xi \in X$ are independent. For example when $X = \mathbb{R}^n$ and $v > 1/n$, the set $\mathscr{S}_v(\mathbb{R}^n)$ of simultaneously v-approximable points is null, while when $v > n$ the set $\mathscr{L}_v(\mathbb{R}^n)$ of dually v-approximable points is null, so that the set of very well approximable points is null in $\mathbb{R}^n$. One of our objectives is to establish Khintchine type results when $X \subset \mathbb{R}^n$ is a manifold, such as a curve or surface, of dimension m less than n. First, the relevant properties of Euclidean submanifolds are summarised.

1.4. Euclidean submanifolds

The manifolds M we consider are smooth m-dimensional immersed submanifolds of Euclidean space $\mathbb{R}^n$ and thus are curves, surfaces and so on (possibly self-intersecting). As usual, they will be taken to be without boundary. They will usually satisfy some analytic or geometric conditions which prevent them having any significant 'flat' regions. We shall also consider only manifolds with no relatively open subsets contained in an $(n-1)$-dimensional affine space (or in geometric terminology 'linearly full' manifolds). As we are concerned with the measure of subsets of manifolds, we can make the important simplification of working locally on the manifold. In turn this means that we can assume that M is an *embedded* manifold (for terminology and further details, see [62], [112], [163], [206]).

To set notation, some formal definitions are given. Let $V \subset \mathbb{R}^n$ and $W \subset \mathbb{R}^m$ be open sets. A function $f\colon V \to U$ is called C^r or *smooth* partial derivatives $\partial^r f/\partial x_{j_1} \dots \partial x_{j_r}$ exist and are continuous. Given sets $X \subset \mathbb{R}^n$ and $Y \subset \mathbb{R}^m$, the function $f\colon X \to Y$ is smooth if for each $x \in X$ there exist an open neighbourhood $V \subset \mathbb{R}^n$ of x and a smooth function $F\colon V \to \mathbb{R}^m$ that coincides with f on $V \cap X$. The function $f\colon X \to Y$ is a *diffeomorphism* if f takes X homeomorphically onto Y and if f and f^{-1} are smooth; X and Y are then called diffeomorphic.

1.4.1. Local parametrisations. A subset $M \subset \mathbb{R}^n$ is a *smooth manifold of dimension* m if M is locally diffeomorphic to $\mathbb{R}^m$. Thus for each point $x \in M$, there are a neighbourhood $V \subset \mathbb{R}^n$ of x and a smooth map $g\colon V \to \mathbb{R}^m$ such that the restriction of h to the submanifold or *patch* $V \cap M$ is a diffeomorphism onto an open subset U of $\mathbb{R}^m$ with smooth inverse map $\theta\colon U \to V \cap M$. The restriction $h\colon V \cap M \to U$ is called a *(local) chart* about ξ and $\theta\colon U \to V \cap M$ is called a *(local) parametrisation* (or coordinate system) of $V \cap M$; the patch $V \cap M$ will be denoted by M_U. The open set $U = h(V \cap M)$ is called a parametrisation domain and we will denote the choice of function and domain by (θ, U). The embedding of the manifold in Euclidean space $\mathbb{R}^n$ determines the coordinates up to relabelling.

It is clear that there is no loss of generality if V is chosen to be a subset of a unit cube, so that $\ell(V) \leqslant 1$. Moreover the nature of the Diophantine inequalities considered here allows translation by integer vectors. For let $\xi \in M$, $\xi' = \xi - \mathbf{r}$ where $\mathbf{r} \in \mathbb{Z}^n$; and let $q \in \mathbb{N}$, $\mathbf{q} \in \mathbb{Z}^n$, $\mathbf{p}' = \mathbf{p} - \mathbf{q}r$. Then

$$q\xi - \mathbf{p} = q\xi' - \mathbf{p}'.$$

Similarly if $p \in \mathbb{Z}$, $\mathbf{q} \in \mathbb{Z}^n$, then $\mathbf{q}\cdot\xi - p = \mathbf{q}\cdot\xi' - p'$ where $p' = p - \mathbf{q}\cdot\mathbf{r}$. Thus the inequalities $|\langle q\xi\rangle| < \psi(q)$ and $\|\mathbf{q}\cdot\xi\| < \psi(|\mathbf{q}|)$ hold for $\xi \in M$ if and only if $|\langle q\xi'\rangle| < \psi(q)$ and $\|\mathbf{q}\cdot\xi'\| < \psi(|\mathbf{q}|)$ respectively hold for $\xi' \in M - \mathbf{r}$. One can also choose $U \subseteq [-1,1]^m$. It follows that we can take $M_U \subset [-1,1]^n$ without loss of generality. From now on these choices for U and M_U will be made when appropriate and usually tacitly.

1.4.2. Monge domains and patches. The parametrisation (θ, U) can be modified to the following convenient form. Let $\widetilde{u} \in U$ and $\widetilde{\xi} = \theta(\widetilde{u}) \in M_U$. Then since $\theta\colon U \to M_U$ is a diffeomorphism, the Jacobian of θ at $\widetilde{u}$ is of maximal rank m and so has a non-vanishing $m \times m$ principal minor $(\partial\theta_i(\widetilde{u})/\partial u_j)$, where by relabelling we can choose $i, j = 1, \dots, m$. Let

$$\widehat{\theta}(\widetilde{u}) = (\theta_1(\widetilde{u}), \dots, \theta_m(\widetilde{u})) = (\widetilde{\xi}_1, \dots, \widetilde{\xi}_m) = \pi_{\mathbb{R}^m}(\theta(\widetilde{u})),$$

where $\pi_{\mathbb{R}^m}\colon \mathbb{R}^{m+k} \to \mathbb{R}^m$ is the usual projection. By the Inverse function theorem, there exists a neighbourhood $U' \subset U$ of $\widetilde{u}$ such that $u = \widehat{\theta}^{-1}(\widehat{\xi})$ for each $u \in U'$, where $\widehat{\xi} = (\xi_1, \ldots, \xi_m)$. Hence for each $u \in U'$,

$$\begin{aligned}\theta(u) &= (\widehat{\theta}(u), \theta_{m+1}(u), \ldots, \theta_n(u)) \\ &= ((\widehat{\theta} \circ \widehat{\theta}^{-1})(\widehat{\xi}), (\theta_{m+1} \circ \widehat{\theta}^{-1})(\widehat{\xi}), \ldots, (\theta_n \circ \widehat{\theta})^{-1})(\widehat{\xi})) \\ &= (\widehat{\xi}, \varphi(\widehat{\xi}))\end{aligned}$$

where $\varphi \colon \widehat{\theta}(U') \to \mathbb{R}^k$ ($k = n - m$, the codimension of M) is a smooth function given by $\varphi_j = \theta_{m+j} \circ \widehat{\theta}^{-1}$ for each $j = 1, \ldots, k$. Thus we can regard $M_{U'} = \widehat{\theta}(U')$ as the graph of φ. From now on we will identify $\widehat{\theta}(U')$ with U' and drop the prime. This parametrisation domain U will be called a *Monge* parametrisation domain and θ a *Monge* parametrisation of the manifold with *ordinate* function φ. Note that the patch

$$M_U = \{\theta(u) \colon u \in U\} = \{(u, \varphi(u)) \colon u \in U\}$$

is the graph of $\varphi \colon U \to \mathbb{R}^k$ and

$$\theta = 1_U \times \varphi. \tag{1.20}$$

The corresponding local chart $h \colon M_U \to U$ is the restriction to M_U of the smooth projection $\mathbb{R}^m \times \mathbb{R}^k \to \mathbb{R}^m$. This is the parametrisation we shall usually (but not invariably) adopt.

By shrinking and closing U if necessary, the existence of the inverse function implies that for each $u \in U$, $i = 1, \ldots, m$, $j = 1, \ldots, k$, we can assume that $|\partial\phi_j / \partial u_i| \leqslant K_{ij}$, where $0 \leqslant K_{ij} < \infty$. Indeed given $\delta > 0$, we can choose U so that for any $u \in U$,

$$K_{ij} - \delta \leqslant |\partial\phi_j(u) / \partial u_i| \leqslant K_{ij}. \tag{1.21}$$

Let

$$K = \max\{K_{ij} \colon i = 1, \ldots, m,\ j = 1, \ldots, k\}. \tag{1.22}$$

Then since $\theta = 1_U \times \varphi$, it follows that for each $i = 1, \ldots, m$, $j = 1, \ldots, n$,

$$|\partial\theta_j(u) / \partial u_i| \leqslant \max\{1, K\}.$$

Similarly when φ and θ are C^r, we can assume that the first r derivatives of θ and φ are bounded on U.

Since M_U is the graph of φ and since U has been chosen suitably, by (1.21) the vector $(q_1, \ldots, q_m, 0, \ldots, 0)$ is not close to being normal to U. This in turn implies that any hyperplane H normal to $(q_1, \ldots, q_m, 0, \ldots, 0)$ meets M_U transversally. It also implies that $H \cap M_U$ is a connected submanifold of M_U of dimension $m - 1$ and is approximately a part of an $(m-1)$-dimensional plane in $\mathbb{R}^n$.

Explicit reference to the manifold M is often omitted and sets of points

$$\{(f_1(x), \ldots, f_n(x)) \colon x \in U\} \text{ or } \{(x_1, \ldots, x_m, f_1(x), \ldots, f_k(x)) \colon x \in U\}$$

with functionally related coordinates can be considered; these coordinates are of course the parametrisation functions for the manifold over U. The question of the orientability of a manifold does not arise since we always work locally. Throughout this tract the manifolds will be at least C^1 and usually C^3. An analytic manifold has analytic local coordinate functions.

1.4.3. The Riemannian metric. The induced Riemannian metric on an embedded manifold in Euclidean space is given locally by

$$ds^2 = \sum_{i,j=1}^{m} g_{ij} du_i\, du_j = \sum_{i,j=1}^{m} \frac{\partial\theta(u)}{\partial u_i} \cdot \frac{\partial\theta(u)}{\partial u_j}\, du_i\, du_j, \tag{1.23}$$

where the metric matrix $G = (g_{ij})$ is symmetric and positive definite and is called the first fundamental form [206, 231]. If the Monge parametrisation is chosen, the metric (1.23) becomes

$$ds^2 = \sum_{i=1}^{m} du_i^2 + \sum_{i,j=1}^{m} \frac{\partial\varphi(u)}{\partial u_i} \cdot \frac{\partial\varphi(u)}{\partial u_j}\, du_i\, du_j. \tag{1.24}$$

1.4.4. Curves. Let I be an interval and let $\theta_j\colon I \to \mathbb{R}$, $j = 1, \ldots, n$, be a family of smooth functions. A smooth curve $\Gamma = \theta(I)$ in n-dimensional Euclidean space is the collection of points

$$(\theta_1(t), \ldots, \theta_n(t)) = \theta(t), t \in I.$$

Here $\theta\colon I \to \mathbb{R}^n$ is a parametrisation of the curve. Curves play an important role both as manifolds in their own right and as geodesics on a manifold. The distance between two points on a manifold is defined to be the length of the geodesic joining them. However, since we will be working locally, we can consider points $\xi, \xi' \in M$ sufficiently near to each other so that the distance $|\xi - \xi'|$ between them in the sup metric is comparable to the distance in the Riemannian metric. As usual, we make a distinction between a curve Γ and a parametrisation (or path) θ.

For a suitable interval I, the parametrisation $\theta(u) = (\theta_1(u), \ldots, \theta_n(u))$ of the curve can be expressed locally in Monge form as $\{(u, \varphi(u))\colon u \in I\}$. The Monge parametrisation is not always adopted. Indeed if the curve is regular (*i.e.*, has non-vanishing tangent), it is often more appropriate to parametrise the curve by arc-length (see [16, 198]). For example, the set

$$\mathbb{S}^+ = \{(x, (1 - x^2)^{1/2})\colon x \in (-1, 1)\}$$

consisting of the upper half of the unit circle $\mathbb{S}^1$ is a Monge domain with ordinate function $\varphi\colon (-1, 1) \to \mathbb{R}$ given by

$$\varphi(x) = (1 - x^2)^{1/2},$$

while using the arc-length s, $\mathbb{S}^+$ has the simple parametrisation

$$\vartheta\colon (0, \pi) \to \mathbb{S}^1\colon s \mapsto (\cos s, \sin s).$$

The rational normal curve $\mathscr{V}$ given by (1.6) is in Monge form. It is closely related to the distribution of the values of integer polynomials and plays an important part in the theory (see §1.5.3).

1.4.5. Bi-Lipschitz parametrisations. The manifolds considered will be assumed to be C^r, where usually $r \geqslant 2$. Each domain U will be chosen so that the local parametrisations $\theta\colon U \to M_U$ are C^r diffeomorphisms with derivatives up to r-th order bounded on U. In addition, each U will be chosen so that θ is bi-Lipschitz on U, *i.e.*, there exist positive constants c, c' such that for each $u, u' \in U$,

$$c'|u - u'| \leqslant |\theta(u) - \theta(u')| \leqslant c|u - u'|.$$

As the local chart h is the inverse of θ, h is bi-Lipschitz on M_U.

For later applications this decomposition will be taken a stage further. Each domain U will be decomposed into countably many hypercubes U_0 say where for each $u_0 \in U_0$,

$$U_0 = \{u \in U\colon |u - u_0| < \varepsilon_0\}$$

of side-length $2\varepsilon_0$. By passing to the Monge form of the parametrisation where $\theta = 1_{U_0} \times \varphi$, we see that the manifold M can be decomposed into countably many graphs M_{U_0} of the smooth ordinate maps $\varphi\colon U_0 \to \mathbb{R}^k$ with derivatives up to r-th order bounded. Since $\theta = 1 \times \varphi$, the inequality $|\theta(u) - \theta(u')| \geqslant |u - u'|$ holds for each $u, u' \in U_0$, so that $c' = 1$. Thus, dropping the suffix 0, the manifold M can be decomposed into countably many subsets M_U, *i.e.*,

$$M = \bigcup_U M_U, \tag{1.25}$$

where the union is over countably many, sufficiently small hypercubes for which the parametrisation θ is bi-Lipschitz on U. As well, the ordinate function φ and its first r derivatives are bounded on U by a constant depending on U and r. Thus M can be viewed locally as the graph of a smooth function (in m independent variables) defined on a suitable open hypercube in $\mathbb{R}^m$. This decomposition into the sets M_U provides a natural framework for studying Lebesgue measure and Hausdorff dimension of subsets of a manifold and will be used a great deal in what follows.

1.4.6. Induced measure on submanifolds. The subset A of the manifold M has induced Lebesgue measure $|A|_M$ given by

$$|A|_M = \int_M \chi_A(\xi)\, dV$$

where dV is the volume element for the induced Riemannian metric (1.23) on M and is given locally by $dV = (\det G)^{1/2}\, du_1 \ldots du_m$ (more details are in [62], [206]). The measure is determined by the Riemannian matrix G. For example when the manifold is the planar curve Γ given by

$$\Gamma = \{x(t) = (x_1(t), x_2(t))\colon t \in \mathbb{R}\},$$

the induced Lebesgue measure of the arc A joining the point $x(t_0)$ to the point $x(t_1)$ is

$$|A|_\Gamma = \int_\Gamma \chi_A(x)\, ds = \int_A ds = \int_{t_0}^{t_1} (x_1'(t)^2 + x_2'(t)^2)^{1/2} dt,$$

which is the arc-length of Γ between the points $x(t_0)$ and $x(t_1)$.

1.4.7. Curvature. This geometric property plays an important part in the study of Diophantine approximation on manifolds. For a planar curve

$$\Gamma = \{x(s) : s \in I\} = \{(x_1(s), x_2(s)) : s \in I\},$$

where the parameter s is the arc-length, the curvature κ is defined to be the derivative of the angle of the tangent at the point $x(s)$. Since $ds^2 = dx_1^2 + dx_2^2$, we have

$$x_1'(s)^2 + x_2'(s)^2 = 1$$

and $(x_1'(s), x_2'(s)) = e^{2\pi i\alpha(s)} \in \mathbb{S}^1$, where $\alpha(s)$ is the angle of the tangent $x'(s)$ to the curve Γ at $x(s)$. Thus $x_1'(s) = \cos\alpha(s)$ and $x_2'(s) = \sin\alpha(s)$, whence

$$\alpha'(s) = x_1'(s)x_2''(s) - x_1''(s)x_2'(s) = \kappa(s).$$

The curvature of a planar curve is somewhat special and $\kappa(s)$ changes sign if the sign of s is reversed, corresponding to reversing the orientation. The zeros of the curvature κ are called points of inflexion and when $\kappa(s) = 0$ in an interval I, the curve $\{x(s) : s \in I\}$ is a straight line segment. When the curve is C^3, κ is a continuously differentiable function of s. In $\mathbb{R}^3$ and higher dimensional Euclidean spaces, the curvature $\kappa(s)$ is defined to be the length of the derivative of the unit tangent vector with respect to arc-length; as such it is non-negative and independent of orientation [231]. The Wronskian $(d^i\theta_j(u)/du^i)$ of the C^n curve $\{\theta(u) : u \in (a,b)\}$ is useful for analysing curves (see [25], [198]) and when $n = 2$ is essentially the curvature.

For manifolds of higher dimensions, more sophisticated differential geometry is needed. The intrepid reader will find the details in Chapter IX, Volume 2 of [140] but others might prefer the following more elementary account. For any $\xi \in M$, let $\theta : U \to M$ be a parametrisation and let $T_\xi M$ and $T_\xi M^\perp$ denote the tangent and normal spaces respectively of M at $\xi = \theta(u)$. The vector space $T_x M$ has a basis

$$X_i = \frac{d}{dt}\Big|_{t=0} \theta(u_1, \ldots, u_i + t, \ldots, u_m) = \frac{\partial\theta}{\partial u_i}, \quad i = 1, \ldots, m.$$

The derivative of X_i in the direction X_j is denoted by $dX_i(X_j)$ and is given by

$$\begin{aligned} dX_i(X_j) &= \frac{d}{ds}\Big|_{s=0} X_i(u_1, \ldots, u_j + s, \ldots, u_m) \\ &= \frac{d}{ds}\Big|_{s=0} \frac{d}{dt}\Big|_{t=0} \theta(u_1, \ldots, u_i + t, \ldots, u_j + s, \ldots, u_m) \\ &= \frac{\partial^2 \theta}{\partial u_i \partial u_j} = \frac{\partial X_i}{\partial u_j} = dX_j(X_i). \end{aligned}$$

This vector need not be tangential to M. Let

$$V = \sum_{i=1}^m V_i X_i, \quad W = \sum_{i=1}^m W_i X_i$$

be two tangent vectors to M at ξ (*i.e.*, $V, W \in T_\xi M$). Extend w, as we may, to a local vector field on M in a neighbourhood of ξ. Then by Leibniz' rule,

$$\begin{aligned} dW(V) &= \sum_{i,j=1}^m V_i d(W_j X_j)(X_i) = \sum_{i,j=1}^m \left(V_i \frac{\partial W_j}{\partial u_i} X_j + W_j V_i dX_i(X_j) \right) \\ &= \sum_{i,j=1}^m V_i \frac{\partial W_j}{\partial u_i} X_j + \sum_{i,j=1}^m V_i W_j \frac{\partial^2 \theta}{\partial u_i \partial u_j}. \end{aligned}$$

The first sum is a linear combination of the X_j and so lies in the tangent space $T_\xi M$. The normal component $dW(V)^\perp$ of $dW(V)$ is therefore

$$dW(V)^\perp = \sum_{i,j=1}^m V_i W_j \frac{\partial^2 \theta}{\partial u_i \partial u_j}.$$

It is evident (and remarkable) that this is both independent of the chosen vector field extension of w and symmetric in V, W. The vector valued symmetric bilinear form $a \colon T_\xi M \times T_\xi M \to T_\xi M^\perp$ given by $a(V, W) = (dW(V))^\perp$ is called the *second fundamental form of* M at $\xi \in M$ and contains information about the movement of the manifold away from the tangent space. Given a unit (in the $|\cdot|_2$ norm) normal vector ν, the second fundamental form of M with respect to ν is the mapping $a_\nu \colon T_\xi M \times T_\xi M \to \mathbb{R}$ given by

$$a_\nu(V, W) = \nu \cdot a(V, W) = \nu \cdot dW(V)^\perp.$$

For fixed ν, a_ν is a bilinear and symmetric form on $T_\xi M$.

We now specialise the parametrisation to the (Monge) form $\theta = 1_U \times \varphi$. Then the basis vectors X_i, $i = 1, \ldots, m$, of $T_x M$ assume the form

$$X_i = \left(0, \ldots, 1, \ldots, 0, \frac{\partial \varphi_1}{\partial u_i}, \ldots, \frac{\partial \varphi_k}{\partial u_i}\right) = \left(e_i, \frac{\partial \varphi}{\partial u_i}\right),$$

where $e_1, \ldots, e_m$ are the standard basis vectors for $\mathbb{R}^m$, and

$$dX_i(X_j) = \frac{\partial^2 \theta}{\partial u_i \partial u_j} = \left(0, \frac{\partial^2 \varphi}{\partial u_i \partial u_j}\right) \in \mathbb{R}^m \times \mathbb{R}^k.$$

Moreover, as is readily verified, the vectors

$$Y_j = \left(\frac{\partial \varphi_j}{\partial u_1}, \dots, \frac{\partial \varphi_j}{\partial u_m}, 0, \dots, -1, \dots, 0\right) = (\operatorname{grad} \varphi_j, -e_j),$$

where $j = 1, \dots, k,$, form a basis of $T_\nu M^\perp$ (here $e_1, \dots, e_k$ are the standard basis vectors for $\mathbb{R}^k$). Let $\nu = \sum_{r=1}^k a_r Y_r$, $a_r \in \mathbb{R}$, be a unit vector in $T_x M^\perp$ and let $\widetilde{\nu} = -\sum_{r=1}^k a_r e_r \in \mathbb{R}^k$. Then

$$\begin{aligned} a_\nu(V, W) = \nu \cdot \left(\sum_{i,j=1}^m V_i W_j \frac{\partial^2 \theta}{\partial u_i \partial u_j}\right) &= \sum_{i,j=1}^m V_i W_j \sum_{r=1}^k a_r Y_r \cdot \left(0, \frac{\partial^2 \varphi}{\partial u_i \partial u_j}\right) \\ &= \sum_{i,j=1}^m \left(-\sum_{r=1}^k a_r \frac{\partial^2 \varphi_r}{\partial u_i \partial u_j}\right) V_i W_j = \sum_{i,j=1}^m \left(\widetilde{\nu} \cdot \frac{\partial^2 \varphi}{\partial u_i \partial u_j}\right) V_i W_j. \end{aligned}$$

The bilinear form $a_\nu \colon T_\xi M \times T_\xi M \to \mathbb{R}$ induces a linear map a from $T_\xi M$ to $(T_\xi M)^*$, the dual space with basis $\{X_1^*, \dots, X_m^*\}$ defined by $X_i^*(X_j) = \delta_{ij}$, given by $a(V) = a_\nu(V, \cdot)$. The matrix A of a with respect to the bases $\{X_1, \dots, X_m\}$ and $\{X_1^*, \dots, X_m^*\}$ is given by the Hessian of $(\widetilde{\nu} \cdot \varphi)$, *i.e.*, $A = \operatorname{Hess}(\widetilde{\nu} \cdot \varphi)$.

The bilinear form $\mathcal{G}(V, W) = \sum_{i,j=1}^m g_{ij} V_i W_j$ formed from the metric (1.23) or (1.24) induces the linear map $g \colon T_x M \to (T_x M)^*$ given by $g(V) = \mathcal{G}(V, \cdot)$. The map g is represented by the matrix $G = (g_{ij})$ with respect to the bases $\{X_1, \dots, X_m\}$ and $\{X_1^*, \dots, X_m^*\}$.

The composition $g^{-1}a \colon T_x M \to T_x M$ is called the *shape operator* and will be denoted by $\mathcal{A}$ $(= \mathcal{A}(\xi, \nu))$. It is symmetric and its eigenvalues $\kappa_i = \kappa_i(x, \nu)$, $i = 1, \dots, m$, are therefore real. The eigenvalues κ_i are the *principal curvatures of M at x with respect to ν*; the eigenvectors are the principal directions. The eigenvalues κ_i are also the eigenvalues of the matrix $G^{-1}A$, so that the principal curvatures of M at x with respect to ν are given by the solutions of the equation

$$\det(G^{-1}A - \kappa 1) = \det G^{-1} \det(A - \kappa G) = 0.$$

Since A and G are symmetric and G is positive definite, A and G can be simultaneously diagonalised. It follows that the eigenvalues α_i of $\mathcal{A}$ are given by

$$\alpha_i = \gamma_i \kappa_i, \ i = 1, \dots, m, \tag{1.26}$$

where $\gamma_1, \dots, \gamma_m$ are the (positive) eigenvalues of the metric matrix G. Thus the eigenvalues of the matrix

$$A = \operatorname{Hess}(\widetilde{\nu} \cdot \varphi) = -\left(\sum_{r=1}^k a_r \frac{\partial^2 \varphi_r}{\partial u_i \partial u_s}\right)$$

have the same sign as the principal curvatures. When its principal curvatures vanish throughout U, the manifold M_U is flat and is a subset of an m-dimensional plane embedded in $\mathbb{R}^n$.

1.5. Metric Diophantine approximation on manifolds

We are now able to discuss metric Diophantine approximation on smooth manifolds embedded in $\mathbb{R}^n$. The restriction to manifolds has two consequences. First if the dimension of the manifold M is strictly less than n, then M and all its subsets are null sets. This presents no difficulty for Hausdorff dimension, which was designed for the study of such sets (see Chapter 3). The appropriate course for Lebesgue measure is to work relative to the induced measure, so that the analysis is in terms of 'almost all' or 'almost no' points on the manifold. Secondly and more significantly, in contrast with a point in Euclidean space, the coordinates (or variables) of a point $\xi = (\xi_1, \ldots, \xi_n)$ on a manifold are functionally related or *dependent*. When non-linear, these constraints pose real difficulties and are the reason why the metric theory is in a far less developed state in the case of dependent variables than for independent ones. Nevertheless significant progress has been made over the last two decades. For example, Khintchine's result that the set of badly approximable points on the rational normal curve $\mathscr{V}$ is relatively null [136] was extended by R. C. Baker to C^1 manifolds with Jacobian having maximal rank almost everywhere [14].

One objective which we achieve only in part is to establish Khintchine's theorem for simultaneous Diophantine approximation and its dual (Corollary 1.4) for points in a smooth submanifold M embedded in $\mathbb{R}^n$. Thus we relate the induced Lebesgue measure of the sets

$$\mathscr{S}(M;\psi) = \{\xi \in M \colon |\langle q\xi \rangle| < \psi(q) \text{ for infinitely many } q \in \mathbb{N}\}$$

and

$$\mathscr{L}(M;\psi) = \{\xi \in M \colon \|\mathbf{q}\cdot\xi\| < \psi(|\mathbf{q}|) \text{ for infinitely many } \mathbf{q} \in \mathbb{Z}^n\}$$

(see §1.3.2) to the convergence or divergence of $\sum_r \psi(r)^n$ and of $\sum_r r^{n-1}\psi(r)$ respectively. The convergence case is established for the set $\mathfrak{M}(\psi)$ (defined in §1.5.3 below and related to the set $\mathscr{L}(\mathscr{V};\psi)$ in Chapter 2). For the sets $\mathscr{S}(M;\psi)$ and $\mathscr{L}(M;\psi)$, the analysis is carried out using local auxiliary sets. Another objective, again only partly realised, is to determine the Hausdorff dimension of $\mathscr{S}(M;\psi)$ and $\mathscr{L}(M;\psi)$ and related sets (see Chapters 4 and 5). The metric structure of $\mathscr{S}(M;\psi)$ and $\mathscr{L}(M;\psi)$ depends on the geometries associated with the 'zero' sets $\{\xi \in \mathbb{R}^n \colon \langle q\xi\rangle = 0\} = \mathbb{Z}^n/q$ and $\{\xi \in \mathbb{R}^n \colon \langle \mathbf{q}\cdot\xi\rangle = 0\}$ of the two dual inequalities (1.7) and (1.9) (see Figure 1.1).

1.5.1. Parametrisation and measure. The decomposition (1.25) is used to analyse the induced measure of the sets $\mathscr{S}(M;\psi)$ and $\mathscr{L}(M;\psi)$. (It is also used for their Hausdorff dimension, as will be seen in Chapter 3 below.) It follows from the discussion in §1.4.1 above, that in the decomposition, the domains U can be chosen so that the parametrisations $\theta \colon U \to M_U$ are bi-Lipschitz on U or equivalently the charts $h \colon M_U \to U$ are bi-Lipschitz on M_U. Consequently the induced Lebesgue measure $|X|_M$ of any subset X of M and the Lebesgue measure

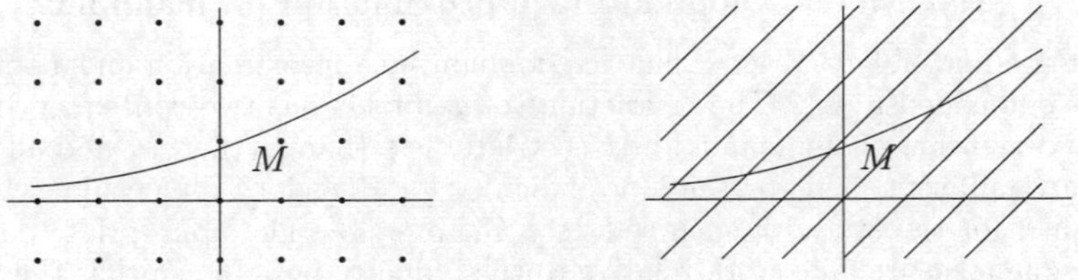

FIGURE 1.1. The manifold M and the two types of 'zero' sets

$|h(X)|_{\mathbb{R}^m}$ are comparable. Indeed let $\{V_j\}$ be a cover by open sets in $\mathbb{R}^n$ of X. Then $\{h(V_j \cap M_U)\}$ is a cover of $h(X \cap M_U)$ satisfying

$$\operatorname{diam}(h(V_j \cap M_U)) \ll \operatorname{diam}(V_j \cap M_U) \ll \operatorname{diam} V_j$$

($\operatorname{diam}(V_j \cap M_U) = \sup\{|x-y| \colon x, y \in V_j \cap M_U\}$ is comparable to the diameter of $V_j \cap M_U$ in the Riemannian metric). It follows that $|h(X \cap M_U)|_{\mathbb{R}^m} \ll |X \cap M_U|_M$ for each U, whence $|h(X)|_{\mathbb{R}^m} \ll |X|_M$. Similarly $|X|_M \ll |h(X)|_{\mathbb{R}^m}$, whence $|X|_M \asymp |h(X)|_{\mathbb{R}^m}$. Thus the study of relatively null subsets (and as we shall see in Chapter 3 of their Hausdorff dimension) of the manifold M fits in very nicely with the manifold's local structure and can be carried out in parameter space. Indeed it follows from the definition of Lebesgue measure that a set $A \subset M$ which can be expressed as a countable union of the form $A = \bigcup_j A_j$ is null if and only if $|A_j|_M = 0$ for each j. Since there are countably many domains U, to prove a subset X of M null, it suffices to prove that each $M_U \cap X$ is null. In particular, $|\mathscr{S}(M;\psi)|_M = 0$ if and only if each $|\mathscr{S}(M_U;\psi)|_M = 0$; and $|\mathscr{L}(M;\psi)|_M = 0$ if and only if each $|\mathscr{L}(M_U;\psi)|_M = 0$.

1.5.2. The auxiliary sets $S(\psi)$ and $L(\psi)$. The metric structure of $\mathscr{S}(M;\psi)$ and $\mathscr{L}(M;\psi)$ can be analysed locally in parameter space by means of the bi-Lipschitz chart $h \colon M_U \to U$. Let $S(\psi) = h(\mathscr{S}(M_U;\psi))$, so that

$$S(\psi) = \{u \in U \colon |\langle q\theta(u)\rangle| < \psi(q) \text{ for infinitely many } q \in \mathbb{N}\}, \tag{1.27}$$

and let $L(\psi) = h(\mathscr{L}(M_U;\psi))$, so that

$$L(\psi) = \{u \in U \colon \|\mathbf{q}\cdot\theta(u)\| < \psi(|\mathbf{q}|) \text{ for infinitely many } \mathbf{q} \in \mathbb{Z}\}. \tag{1.28}$$

Then by the preceding sub-section,

$$|S(\psi)|_{\mathbb{R}^m} \asymp |\mathscr{S}(M_U;\psi)|_M \tag{1.29}$$

and

$$|L(\psi)|_{\mathbb{R}^m} \asymp |\mathscr{L}(M_U;\psi)|_M. \tag{1.30}$$

Hence the sets $\mathscr{S}(M;\psi)$ and $\mathscr{L}(M;\psi)$ are null if and only if $S(M;\psi)$ and $L(M;\psi)$ respectively are null for each U. When discussing $S(\psi)$ and $L(\psi)$, the patches M_U and domains U will usually be taken to be in $[-1,1]^n$ and $[-1,1]^m$ respectively.

These sets can be put into a 'lim-sup' form by expressing the inequality in terms of inclusion in the appropriate set. Given any positive $\delta < 1/2$, $q \in \mathbb{N}$ and $\mathbf{p} \in \mathbb{Z}^n$, let

$$B_\delta(\mathbf{p}, q) = \{u \in U \colon |q\,\theta(u) - \mathbf{p}| < \delta\} \tag{1.31}$$

and let

$$B_\delta(q) = \bigcup_{\mathbf{p}} B_\delta(\mathbf{p}, q) = \{u \in U \colon |\langle q\,\theta(u)\rangle| < \delta\},$$

where the union is over $\mathbf{p} \in \mathbb{Z}^n$ with $|q\theta(u) - \mathbf{p}| < \delta$ for some u (for each $u \in U$, $q \in \mathbb{N}$, there exists at most one such $\mathbf{p}$). Then

$$S(\psi) = \{u \in U \colon u \in B_{\psi(q)}(q) \text{ for infinitely many } q \in \mathbb{N}\}. \tag{1.32}$$

Similarly

$$L(\psi) = \{u \in U \colon u \in B_{\psi(|\mathbf{q}|)}(\mathbf{q}) \text{ for infinitely many } \mathbf{q} \in \mathbb{Z}^n\}, \tag{1.33}$$

where

$$B_\delta(\mathbf{q}) = \bigcup_{p} B_\delta(p, \mathbf{q}) = \{u \in U \colon |\langle \mathbf{q} \cdot \theta(u)\rangle| < \delta\},$$

where the union is over integers p such that $|\mathbf{q} \cdot \theta(u) - p| < \delta$ and

$$B_\delta(p, \mathbf{q}) = \{u \in U \colon |\mathbf{q} \cdot \theta(u) - p| < \delta\} = \{u \in U \colon |\mathbf{q} \cdot (u, \varphi(u)) - p| < \delta\}.$$

Thus if for each parametrisation domain U the sum

$$\sum_{q=1}^{\infty} |B_{\psi(q)}(q)| < \infty, \tag{1.34}$$

then by Cantelli's lemma, $|S(\psi)|_{\mathbb{R}^m} = 0$. But $|S(\psi)|_{\mathbb{R}^m} \asymp |\mathscr{S}(M_U; \psi)|_M$ by (1.29), whence $|\mathscr{S}(M_U; \psi)|_M = 0$ for each U and so $|\mathscr{S}(M\,; \psi)|_M = 0$. Hence the proof that $|\mathscr{S}(M\,; \psi)|_M = 0$ reduces to establishing (1.34).

Similarly, in the linear forms case, to prove that $|\mathscr{L}(M\,; \psi)|_M = 0$ reduces to establishing for each parametrisation domain U that

$$\sum_{\mathbf{q} \in \mathbb{Z}^n \setminus \{0\}} |B_{\psi(|\mathbf{q}|)}(\mathbf{q})| < \infty.$$

By Cantelli's lemma, this implies that $|L(\psi)|_{\mathbb{R}^m} = 0$, which in turn implies by (1.30) that $|\mathscr{L}(M_U; \psi)|_M = 0$ and it follows that $|\mathscr{L}(M; \psi)|_M = 0$.

The sets $S(\psi)$ and $L(\psi)$ depend on U and θ but while the smoothness of θ is important, the particular choices of domain and map are not and accordingly this dependence will usually not be expressed. The notation $S(\psi)$ and $L(\psi)$ will not be changed when we pass to the Monge form where $\theta = 1_U \times \varphi$ and

$$q\,\theta(u) = (q\,u, q\,\varphi(u)), \quad \mathbf{q} \cdot \theta(u) = \mathbf{q}^{(1)} \cdot u + \mathbf{q}^{(2)} \cdot \varphi(u),$$

where unless otherwise stated,

$$\mathbf{q}^{(1)} = (q_1, \dots, q_m) \in \mathbb{Z}^m, \quad \mathbf{q}^{(2)} = (q_{m+1}, \dots, q_n) \in \mathbb{Z}^k \tag{1.35}$$

and $\mathbf{q} = (\mathbf{q}^{(1)}, \mathbf{q}^{(2)})$. When $\psi(r) = r^{-v}$, we write S_v for $S(\psi)$ and L_v for $L(\psi)$.

For each $q \in \mathbb{N}$, the set

$$R_q = \{u \in U \colon \langle q\,\theta(u)\rangle = 0\} = \{u \in U : \langle (qu, q\varphi(u))\rangle = 0\},$$

associated with simultaneous Diophantine approximation, is the projection of the set $(\mathbb{Z}^n/q) \cap M$ and the set

$$R_{\mathbf{q}} = \{u \in U : \langle \mathbf{q} \cdot \theta(u)\rangle = 0\} = \{u \in U : \langle \mathbf{q} \cdot (u, \varphi(u))\rangle = 0\},$$

associated with the dual form, is the projection of the set $\{\xi \in M \colon \langle \mathbf{q} \cdot \xi\rangle = 0\}$. They can be regarded as generalisations of rationals with denominator q and will be called *resonant.*

1.5.3. Integral polynomials and the rational normal curve. A natural measure of the size of the polynomial $P(x) = q_0 + q_1 x + \cdots + q_n x^n$ is its *height*

$$h(P) = \max\{|q_0|, \dots, |q_n|\}$$

and it leads to consideration of the inequality

$$|P(t)| < \psi(h(P)) \tag{1.36}$$

rather than $\|\mathbf{q} \cdot (t, \dots, t^n)\| < \psi(|\mathbf{q}|)$, where $\mathbf{q} = (q_1, \dots, q_n)$. The set of t in the interval I such that (1.36) holds for infinitely many integral polynomials P of degree at most n (or equivalently infinitely many $(q_0, q_1, \dots, q_n) \in \mathbb{Z}^{n+1}$) is denoted by $\mathfrak{M}(\psi) = \mathfrak{M}^{(n)}(\psi; I)$,

$$\mathfrak{M}(\psi) = \{t \in I \colon |P(t)| < \psi(h(P)) \text{ for infinitely many } P\},$$

where the dependence on the interval I and the degree n will be omitted unless confusion might arise. This set is closely related to $L(\psi) = \pi_{\mathbb{R}}(\mathscr{L}(\mathscr{V}; \psi))$ and is discussed further in §2.4.

LEMMA 1.5. *Let $\psi \colon \mathbb{N} \to \mathbb{R}^+$ be a decreasing approximation function and let I be a finite interval. Then there exists a $K < 1$ such that*

$$\mathfrak{M}(\psi) \subset L(\psi) \subset \mathfrak{M}(\psi_K)$$

where $\psi_K(r) = \psi(Kr)$.

PROOF. If $t \in \mathfrak{M}(\psi)$ then there are infinitely many $P(x) = q_0 + q_1 x + \cdots + q_n x^n$ in $\mathbb{Z}[x]$ such that

$$|q_0 + q_1 t + \cdots + q_n t^n| = \|q_1 t + \cdots + q_n t^n\| < \psi(h(P)) \leqslant \psi(|\mathbf{q}|),$$

where $\mathbf{q} = (q_1, \dots, q_n)$, whence $t \in L(\psi)$.

On the other hand if $t \in L(\psi)$, then there are infinitely many $q_0, q_1, \dots, q_n \in \mathbb{Z}$ such that

$$|q_0 + q_1 t + \cdots + q_n t^n| = \|q_1 t + \cdots + q_n t^n\| < \psi(|\mathbf{q}|).$$

It follows that $|q_0| < \psi(|\mathbf{q}|) + |q_1 t + \cdots + q_n t^n|$, whence for sufficiently large $\mathbf{q}$

$$|q_0| < 1/2 + n|\mathbf{q}|c(I) < (c(I)n + 1)|\mathbf{q}|,$$

where $c(I)$ is a constant depending on I ($c(I) = \max\{\sup\{|t|^j : t \in I\} : j = 1, n\}$). Thus $h(P) < |\mathbf{q}|/K$, where $K = (c(I) + 1)^{-1} < 1$ and so since ψ is decreasing,

$$|P(t)| < \psi(Kh(P))$$

for infinitely many integral polynomials of degree at most n. □

When $\psi(q) = q^{-v}$, we write $\mathfrak{M}(\psi)$ as $\mathfrak{M}_v$; thus $\mathfrak{M}_v$ is the set $t \in I$ for which the inequality

$$|P(t)| < h(P)^{-v} \tag{1.37}$$

holds for infinitely many integral polynomials P of degree at most n. By the above lemma,

$$\mathfrak{M}_v \subset L_v \subset \mathfrak{M}_{v-\varepsilon} \tag{1.38}$$

for any $\varepsilon > 0$, since given $\varepsilon' > \varepsilon$, $(h(P)/(n+1))^{v-\varepsilon'}(n+1)^{v-\varepsilon'} < h(P)^{v-\varepsilon}$ for $h(P)$ sufficiently large. The set

$$\mathscr{M}_v = \{t \in I : \omega(t) \geqslant v\}$$

where for each real t, $\omega(t) = \omega_n(t) = \sup\{w \in I : t \in \mathfrak{M}_w\}$, is of interest in transcendence theory (see [13] where $M_n(\lambda) = \mathscr{M}^{(n)}_{(n+1)\lambda-1}$ and [208] where $M_n(w)$ is essentially $\mathfrak{M}^{(n)}_w$). As with the related sets $\mathscr{L}_v$ and $\mathcal{L}_v$,

$$\mathfrak{M}_v \subset \mathscr{M}_v \subset \mathfrak{M}_{v-\varepsilon}$$

for any $\varepsilon > 0$. By analogy with §1.3.3, the set of points for which the exponent v is exact is defined by

$$\mathscr{M}'_v = \{t \in I : w(t) = v\}.$$

As a consequence of Khintchine's transference principle (see §1.3.1), the simultaneous Diophantine approximation of the points $t, t^2, \dots, t^n$ and the value of the integral polynomial $P(t)$ of degree at most n are closely related. Indeed by Khintchine's transference principle, if for a given $\varepsilon > 0$, there are infinitely many integral polynomials $P(t) = q_0 + q_1 t + \cdots + q_n t^n$ of degree n such that

$$|P(t)| < h(P)^{-n-\varepsilon},$$

then there are infinitely many positive integers q such that

$$|\langle q(t, \dots, t^n)\rangle| = |\langle (qt, \dots, qt^n)\rangle| < q^{-(1+\varepsilon')/n},$$

where $\varepsilon' > 0$ is comparable to ε; and the reverse result suitably modified also holds.

1.5.4. The parabola. We illustrate some of the above ideas in the simple case when M is the parabola $\{(t, t^2) : t \in \mathbb{R}\}$. In view of the remarks above about working locally, it suffices to consider $L(\psi)$ instead of $\mathscr{L}(M; \psi)$. We take U to be the interval $I = (-1, 1)$, *i.e.*, we consider the set $L(\psi)$ of points $t \in (-1, 1)$ which satisfy

$$|\langle q_1 t + q_2 t^2 \rangle| < \psi(|\mathbf{q}|) \tag{1.39}$$

for infinitely many integer vectors $\mathbf{q} = (q_1, q_2)$. The resonant sets are the zeros of $\langle q_1 t + q_2 t^2 \rangle$ in I and so are the zeros in I of the integral quadratics $q_0 + q_1 t + q_2 t^2$.

The natural cover of $L(\psi)$ consists of the family $\{B_{\psi(|\mathbf{q}|)}(\mathbf{q}) : |\mathbf{q}| = 1, 2, \dots\}$, where for $0 < \delta < 1/2$,

$$B_\delta(\mathbf{q}) = B_\delta(q_1, q_2) = \{t \in I : |\langle q_1 t + q_2 t^2 \rangle| < \delta\} = \bigcup_{\substack{|p| \leqslant |\mathbf{q}| \\ p \in \mathbb{Z}}} B_\delta(p, \mathbf{q})$$

and $B_\delta(p, \mathbf{q}) = \{t \in I : |q_1 t + q_2 t^2 - p| < \delta\}$. In particular, for non-zero $q \in \mathbb{N}$,

$$|B_\delta(0, (q, 0))| = |\{t \in I : |qt| < \delta\}| = 2\delta/q$$

while

$$|B_\delta(0, (0, q))| = |\{t \in I : |qt^2| < \delta\}| = 2(\delta/q)^{1/2}.$$

Thus the length $2(\delta/q)^{1/2}$ of the intercept of the δ-neighbourhood of the tangent to the parabola at the origin with the parabola is about the square root of the length $2\delta/q$ of the intercept of the neighbourhood of the normal there. This dependence of $|B_\delta(p, \mathbf{q})|$ on the direction of $\mathbf{q} = (q_1, q_2)$ and the disproportionate length when the associated resonant set is a tangent means that the bound for $|B_\delta(\mathbf{q})|$ is too large for the cover to converge. Thus Cantelli's lemma cannot be invoked to deduce that the set of points $u \in I$ for which (1.39) holds infinitely often for a suitable approximation function ψ is null. This is an important difference between approximation involving dependent and independent variables. However, because the 'near-tangencies' do not occur too frequently, they can be dealt with in this and more general cases (the parabola is treated in §2.4.4 below; instead of (1.39), the closely related inequality

$$|Q(t)| = |q_0 + q_1 t + q_2 t^2| \leqslant \psi(h(Q))$$

is considered). When the manifold is too 'flat', as in the case of polynomials with many repeated roots, the cover is unavoidably too large and the associated sum diverges. In the case of simultaneous approximation, the set $B_\delta(\mathbf{p}, q)$ (see (1.31)) has measure $|B_\delta(\mathbf{p}, q)| \ll \delta$; the distribution of the rational vectors $\mathbf{p}/q$ near the parabola plays a crucial part.

1.5.5. Conclusion. A common strategy in proving a lim-sup set null is to estimate the measure of the elements in the natural cover and use Cantelli's lemma. Given a parametrisation domain U of the manifold M, the natural covers for the lim-sup sets $S(\psi)$, $L(\psi)$ and $\mathfrak{M}(\psi)$ will be used repeatedly in the analysis of their measures. The local information obtained for U can be lifted to M_U under the bi-Lipschitz parametrisation of M. Since M is a countable union of these local graphs M_U, the information about the measures of $\mathscr{S}(M_U;\psi)$ and $\mathscr{L}(M_U;\psi)$ can be extended to M. In particular, if $S(\psi)$ and $L(\psi)$ are null for each U, then $\mathscr{S}(M;\psi)$ and $\mathscr{L}(M;\psi)$ respectively are also null. The Hausdorff dimension of these null sets will be studied in Chapters 4 and 5. When $\mathbf{q}$ is close to the normal space $T_\xi M^\perp$ at ξ, the sets $B_\delta(\mathbf{q})$ are close to the tangent space $T_\xi M$ at ξ and so have large measure. This can make establishing convergence and applying Cantelli's lemma problematic.

1.6. Notes

§**1.2.1** E. Borel [51] proved that when $\psi(q) = q^{-1-\varepsilon}$, $\varepsilon > 0$, the set of points satisfying (1.2) was null.

§**1.2.2** The terminology *badly approximable* is used to mean *not* very well approximable in [210, p. 67]. It would seem more appropriate to use (as in [89]) *well approximable* instead of *very well approximable* for the set $\cup_{v>1}\mathscr{K}_v$ and to retain very well approximable for the set $\cap_{v>1}\mathscr{K}_v$ which includes numbers of Liouville type. Nevertheless, to avoid confusion we follow current usage [14], [139], [201, p. 61].

The lim-sup set $\mathscr{K}_v(\mathbb{R})$ is a countable intersection of open dense sets and so is non-meagre or second Baire category in $\mathbb{R}$ [133] but paradoxically is null when $v > 1$. Similarly its complement is of full measure but a countable union of closed nowhere dense sets and so meagre (first Baire category). Similar observations hold for related sets (*e.g.*, the set E_σ in §7.2). There is an interesting discussion of general Diophantine approximation from this point of view in [110].

The approximation functions $\Phi\colon [0,\infty) \to \mathbb{R}^+$ defined by H. R. Rüssmann in [191] satisfy further conditions.

§**1.3.2** H. Davenport and Schmidt showed that a sharpening of the simultaneous and dual forms of Dirichlet's theorem in which the approximation $1/N$ is multiplied by a constant $c < 1$ holds for badly approximable numbers [70] and that the same was true for the parabola, for more details see [71]. R. C. Baker extended this to planar curves [15, 16]; later this was proved for smooth manifolds satisfying certain curvature constraints [92].

Approximation of the form $|\mathbf{q}\cdot x| < \psi(|\mathbf{q}|)$ (see [73], [74]) has also been studied when x lies on a manifold [91].

Diophantine approximation involving complex numbers will not be discussed much, more details are to be found in [16], [94], [144], [208]. More generally one could consider the set of matrices A in $M = \prod_{j=1}^n M_j$, where each $M_j \subset \mathbb{R}^m$, such that (1.5) holds for infinitely many $\mathbf{q} \in \mathbb{Z}^m$. On the whole we confine our attention to $\mathscr{S}(M;\psi)$ and $\mathscr{L}(M;\psi)$.

§**1.3.4** Khintchine used continued fractions in [134],[137]; a geometrical proof of the theorem is given in [214, §3]. More general results obtained using ergodic and probabilistic ideas rely in essence on a pairwise quasi-independence generalisation of the divergence case in the Borel-Cantelli Lemma [141]; see also [49], [63], [83], [115]. Schmidt obtained a quantitative refinement of the Khintchine-Groshev theorem by establishing an asymptotic formula [196], [197], see also [115], [210]; they have also been obtained for certain manifolds [28, 29, 95, 210].

The conjecture of Duffin and Schaeffer that if $\sum_q(\psi(q)\varphi(q)/q)$, where φ is Euler's divisor function, diverges, then $\mathscr{K}(\mathbb{R};\psi)$ has full Lebesgue measure is still open (a full discussion is in [115, Chapter 2]) but a higher dimensional result has been proved by A. D. Pollington and R. C. Vaughan [183].

The approximation function $\psi\colon \mathbb{N} \to \mathbb{R}^+$ in the Khintchine-Groshev theorem can be generalised to a function $\Psi\colon \mathbb{Z}^m \to \mathbb{R}^+$ and the set $W([0,1]^{mn};\psi)$ to the set

$$W([0,1]^{mn},\Psi) = \{A \in [0,1]^{mn} \colon |\langle \mathbf{q}A \rangle| < \Psi(\mathbf{q}) \text{ for infinitely many } \mathbf{q} \in \mathbb{Z}^m\}.$$

The sets $B_\delta(\mathbf{q}) = \{A \in [0,1]^{mn} \colon |\langle \mathbf{q}A \rangle| < \delta\}$, where $\mathbf{q} \in \mathbb{Z}^m$, $0 < \delta < 1/2$, satisfy $|B_\delta(\mathbf{q})| = 2^n\delta^n$ and $|B_\delta(\mathbf{q}) \cap B_{\delta'}(\mathbf{q}')| = |B_\delta(\mathbf{q})||B_{\delta'}(\mathbf{q}')|$ (pairwise independence) for $m \geqslant 2$, $\mathbf{q}, \mathbf{q}'$ distinct and primitive, either by Fourier analysis (Chapter 1, §5 [210]) or torus geometry [83]. If $\sum_{\mathbf{q}} \Psi(\mathbf{q})$ converges, then $|W([0,1]^{mn},\Psi)| = 0$ by Cantelli's lemma (see Chapter 1, §6, in [210]); otherwise $|W([0,1]^{mn},\Psi)| = 1$ by [210, Lemma 5] or [115, Lemma 1.5]. Indeed Schmidt's refinement [196], [197] is used in [115] to prove an asymptotic formula (Theorem 9.1).

Inhomogeneous Diophantine approximation is rather different from homogeneous, as exemplified by Kronecker's theorem and Dirichlet's theorem [59]. There is an inhomogeneous 'doubly metric' inhomogeneous Khintchine theorem in [59, Chapter 5] and Schmidt's more general asymptotic formula [197] implies a 'singly metric' inhomogeneous Khintchine theorem. There are few inhomogeneous results for manifolds but see [27].

The sum in Corollary 1.4 (b) can be replaced by the comparable sum $\sum_{\mathbf{q}} \psi(|\mathbf{q}|)$ over all non-zero $\mathbf{q} \in \mathbb{Z}^n$.

§1.4 The boundary ∂M of an m-dimensional smooth manifold $M \subset \mathbb{R}^n$ consists of those points in M which are the images under a local parametrisation of the boundary $\{0\} \times \mathbb{R}^{m-1}$, *i.e.*, points which have a neighbourhood diffeomorphic to an open set in $[0,\infty) \times \mathbb{R}^{m-1}$. Boundaries will not be of any interest in metric questions, since if it exists, the boundary of a smooth m-dimensional manifold will have dimension $m-1$ and so does not affect Lebesgue measure and Hausdorff dimension.

§1.5.2 Of course $L(\psi)$ can be thrown into lim-sup form by considering $\widetilde{B}_\delta(q) = \bigcup_{|\mathbf{q}|=q} B_\delta(\mathbf{q})$.

§1.5.3 Infinitely many irreducible integer polynomials P of degree at most n satisfy (1.37) when $v < \omega(t)$ [232].

CHAPTER 2

Khintchine's and Groshev's theorems for manifolds

2.1. Introduction

Questions of metric Diophantine approximation are now considered for points $\xi = (\xi_1, \dots, \xi_n)$ lying on submanifolds embedded in Euclidean space $\mathbb{R}^n$. In particular, the Khintchine-Groshev theorem (see §1.3) is extended to certain manifolds. Thus the size, expressed in terms of the induced Lebesgue measure, of sets of ψ-approximable points on these manifolds is related to volume sums involving the approximation term. As a prelude, extremal manifolds, which are an important half-way house, will be discussed briefly. Fuller accounts of Mahler's conjecture and extremality are in [208], [210] and [115, Chapter 9].

Recall that under some mild monotonicity restrictions on the approximation function ψ, the Khintchine-Groshev theorem (see §1.3.4) implies that the system of inequalities

$$|\langle q\xi\rangle| = \max\{\|q\,\xi_j\| : j = 1, \dots, n\} < \psi(q) \tag{2.1}$$

holds for infinitely many positive integers q for almost all or almost no x in $\mathbb{R}^n$ according as the sum $\sum_q \psi(q)^n$ diverges or converges; and the dual inequality

$$\|\mathbf{q}\cdot\xi\| < \psi(|\mathbf{q}|) \tag{2.2}$$

holds for infinitely many integer vectors $\mathbf{q} \in \mathbb{Z}^n$ for almost all or almost no ξ in $\mathbb{R}^n$ according as the sum $\sum_q q^{n-1}\psi(q)$ diverges or converges. The solubility of (2.1) and (2.2) when ξ is a point on a smooth m-dimensional Euclidean submanifold M embedded in $\mathbb{R}^n$ is a much more difficult question. We will be concerned with the metric structure of the sets

$$\mathscr{S}(M;\psi) = \{\xi \in M : |\langle q\xi\rangle| < \psi(q) \text{ for infinitely many } q \in \mathbb{N}\}$$

and

$$\mathscr{L}(M;\psi) = \{\xi \in M : \|\mathbf{q}\cdot\xi\| < \psi(|\mathbf{q}|) \text{ for infinitely many } \mathbf{q} \in \mathbb{Z}^n \},$$

introduced in §1.3.2). Much of this chapter is devoted to the proof of a conjecture due to A. Baker concerning $\mathscr{L}(\mathscr{V};\psi)$ (see §2.4).

2.2. Extremal manifolds

A subset M of $\mathbb{R}^n$ is called *extremal* if the set of very well approximable points in M is relatively null. Thus M is extremal if the sets $\mathscr{S}_v(M)$ and $\mathscr{L}_v(M)$ (defined in Chapter 1, §1.3.2) are null in M when $v > 1/n$ and $v > n$ respectively (extremality is not affected by the different resonant geometries). Extremality can be regarded as a step towards the more delicate questions of Khintchine and Groshev type results and of Hausdorff dimension, discussed in §2.3 and Chapters 4 and 5 below. The terminology reflects the fact that for almost all points on an extremal set, the exponents in Corollary 1.3 to Dirichlet's theorem are unimprovable [142, p. 67]; it should not be taken to mean that the property is exceptional since generically speaking smooth manifolds are extremal.

Extremal manifolds arose in connection with K. Mahler's classification of transcendence of real numbers. This led him to conjecture in 1932 that the set S_v of real numbers t such that

$$\max\{\|qt^j\| : j = 1, \dots, n\} = |\langle q(t, \dots, t^n)\rangle| < q^{-v} \tag{2.3}$$

holds for infinitely many $q \in \mathbb{N}$ is null when $v > 1/n$ (he proved that S_v was null for $v > 4/n$) [152]. By Khintchine's transference principle this is equivalent to the set L_v of $t \in \mathbb{R}$ such that

$$\|\sum_{j=1}^{n} q_j t^j\| = |q_0 + q_1 t + \cdots + q_n t^n| < |\mathbf{q}|^{-v} \tag{2.4}$$

holds for infinitely many $\mathbf{q} \in \mathbb{Z}^n$ being null when $v > n$; and by (1.38) to $\mathfrak{M}_v$ (see §1.5.3) being null for $v > n$. By §1.5.2, $\mathscr{S}_v(\mathscr{V})$ is relatively null for $v > 1/n$ and $\mathscr{L}_v(\mathscr{V})$ is relatively null for $v > n$. Thus Mahler's conjecture can be interpreted as the rational normal curve $\mathscr{V} = \{(t, t^2, \dots, t^n) : t \in \mathbb{R}\}$ being extremal.

J. Kubilius proved the conjecture in the case $n = 2$ in 1949, thus showing that the parabola was extremal [145]. In 1964 Schmidt established the striking result that any C^3 planar curve $\Gamma = \{(x_1(s), x_2(s)) : s \in \mathbb{R}\}$ with non-zero curvature almost everywhere is extremal; in our notation $|\mathscr{L}_v(\Gamma)|_\Gamma = 0$ for any $v > 2$ [198] (a proof is given in Chapter 9 of [115]). Sprindžuk's proof [208] (also in the vintage year of 1964) of Mahler's conjecture opened the way to the theory of metric Diophantine approximation of dependent quantities. Sprindžuk conjectured that if the parametrisation maps $\theta_j : U \to \mathbb{R}$, $j = 1, \dots, n$, are analytic and, together with 1, independent over $\mathbb{R}$, then the manifold

$$M = \{(\theta_1(u), \dots, \theta_n(u)) : u \in U\}$$

is extremal (Conjecture H_1 in [211]). A manifold M is *strongly extremal* if given any $v > n$, the set of points $x = (x_1, \dots, x_n) \in M$ satisfying

$$\|\mathbf{q} \cdot x\| < \prod_{j=1}^{n} (|q_j| + 1)^{-v/n}, \tag{2.5}$$

for infinitely many $\mathbf{q} \in \mathbb{Z}^n$ is null in M. A strongly extremal manifold is extremal and any point in $\mathscr{L}_v(M)$ also satisfies (2.5) infinitely often. A. Baker conjectured that the rational normal curve $\mathscr{V}$ is strongly extremal in (2.3) [11] and Sprindžuk extended the conjecture to any manifold M satisfying the conditions of H_1 (Conjecture H_2 in [211]).

Manifolds satisfying a variety of analytic, geometric and number theoretic conditions have been shown to be extremal (more details are in [210], [211], [226]). In an extension of Schmidt's theorem to higher dimensional manifolds, E. Kovalevskaya has shown that surfaces in $\mathbb{R}^3$ having non-zero Gaussian curvature almost everywhere are extremal ([143] or [210], p. 149, Theorem 18) and together with Bernik later extended this result to m-dimensional surfaces in $\mathbb{R}^{2m}$ [40]. These are special cases of the more general result that smooth (C^3) manifolds of dimension at least 2 and satisfying a curvature condition (which specialises to non-zero Gaussian curvature for surfaces in $\mathbb{R}^3$) are also extremal (see [89], [93] and the next section). The rational normal curve $\mathscr{V}$ was shown to be strongly extremal by Bernik [34].

Schmidt's result has been extended to C^4 curves in $\mathbb{R}^3$ by V. V. Beresnevich and Bernik [25]. However, in a remarkable development, D. Y. Kleinbock and G. A. Margulis [139] have proved that manifolds which are nondegenerate almost everywhere are strongly extremal. Nondegeneracy can be regarded as a generalisation of non-zero curvature and is defined as follows. For each $j \leqslant k$, the point $x = \theta(u) \in M \subset \mathbb{R}^n$ is *j-nondegenerate* if the partial derivatives of θ at u up to order j span $\mathbb{R}^n$. The point x is *nondegenerate* if it is *j-nondegenerate* for some j. This result is best possible and implies both Sprindžuk's and the stronger Baker-Sprindžuk conjectures. The proof uses unipotent flows in homogeneous spaces of lattices, the correspondence between multiplicatively very well approximable points and unbounded orbits in the space of lattices. At the moment their techniques do not yield the Hausdorff dimension but they do lead to a generalisation of A. Baker's result [10] and are likely to lead to further progress.

There are extremal lines as well, in all dimensions: indeed the line

$$\{(u, \alpha_2 u + \beta_2, \ldots, \alpha_n u + \beta_n) \colon u \in \mathbb{R}\}$$

where either each α_j or each β_j, $2 \leqslant j \leqslant n$, is badly approximable is extremal [198]. On the other hand if all the β_j vanish and any α_j is rational, the line is not extremal; if $\alpha_2 = p/q$ then

$$\mathbf{q} \cdot (u, p/q, \alpha_3, \ldots, \alpha_n) = 0$$

for $\mathbf{q} = (-rp, rq, 0, \ldots, 0), r \in \mathbb{Z}$, and all points are very well approximable. The same will be true if the vectors $\alpha \in \mathbb{R}^n$ are approximable to exponent strictly exceeding n. There are planes which are not extremal, *e.g.*, every point in a plane containing the origin with rational normal is very well approximable.

2.3. Khintchine and Groshev type manifolds

Some further terminology is needed to discuss the deeper Khintchine type theory. Let $\psi \colon \mathbb{N} \to \mathbb{R}^+$ be a decreasing approximation function. A manifold M embedded in $\mathbb{R}^n$ will be called a *Khintchine type manifold for convergence* if the convergence of the sum

$$\sum_{q=1}^{\infty} \psi(q)^n \tag{2.6}$$

implies that $|\mathscr{S}(M; \psi)|_M = 0$. (The monotonicity condition on ψ is not always required for convergence but is introduced for convenience.) It will be called a

Khintchine type manifold for divergence if the divergence of the sum (2.6) implies that $|M \setminus \mathscr{S}(M;\psi)|_M = 0$.

Similarly, a manifold M embedded in $\mathbb{R}^n$ will be called *a Groshev type manifold for convergence* if the convergence of the sum

$$\sum_{q=1}^{\infty} q^{n-1}\psi(q) \tag{2.7}$$

(or equivalently if $\sum_{\mathbf{q}\neq 0}\psi(|\mathbf{q}|) < \infty$) implies that $|\mathscr{L}(M\,;\psi)|_M = 0$. It will be called a *Groshev type manifold for divergence* if the divergence of the sum (2.7) implies that $|M \setminus \mathscr{L}(M;\psi)|_M = 0$. We conjecture that a C^k manifold which is nondegenerate almost everywhere is of Khintchine-Groshev type (*i.e.*, the sets of the ψ-approximable points are null or full according as the appropriate sums converge and diverge for both simultaneous and dual Diophantine approximation) (see §2.3 in the Notes).

Fairly general classes of Euclidean submanifolds are of Khintchine or Groshev type. The parabola and more generally any C^3 planar curve Γ with non-zero curvature almost everywhere is a Groshev type manifold [26, 37]. This extends Schmidt's extremality result [198] and R. C. Baker's refinement [16] that the convergence of $\sum_q \psi(q)$ implies that $|\mathscr{L}(\Gamma;\psi^2)|_\Gamma = 0$. The parabola is a Khintchine type manifold for convergence since if $\sum_q \psi(q)^2$ converges, then the inequality

$$|\langle q(t,t^2)\rangle| = \max\{\|qt\|, \|qt^2\|\} < \psi\,(q)$$

has infinitely many integer solutions $q > 0$ for almost no t [31]. Less is known for curves in higher dimensions but the important rational normal curve $\mathscr{V}$ is of Groshev type for convergence, thus establishing another conjecture of A. Baker [10] (treated in the next section). Higher dimensional manifolds are better understood. Indeed C^3 manifolds with dimension at least 2 embedded in $\mathbb{R}^n$ which are 2-convex almost everywhere (an explanation of this term is in the next section) are both of Khintchine and Groshev type for convergence (see Theorem 2.20 below). The study of the Hausdorff dimension of the correponding null sets is taken up in Chapters 4 and 5. Manifolds satisfying stronger curvature conditions enjoy some Khintchine-Groshev type properties. Quantitative results are more difficult and restricted but see [29], [95] and [210, Chapter 2, §12].

2.4. Baker's conjecture

In 1966 A. Baker [10] extended Sprindžuk's theorem and showed that if ψ is monotonically decreasing and the sum $\sum_{r=1}^{\infty}\psi(r)$ converges, then the set of $t \in \mathbb{R}$ such that $|P(t)| < \psi(h(P))^n$ holds for infinitely many integral polynomials P of degree n is null (it is discussed fully in [115, Chapter 9]). He also conjectured the following result.

THEOREM 2.1. *Let $\psi(r)$ be any strictly decreasing function which satisfies*

$$\sum_{r=1}^{\infty} r^{n-1}\psi(r) < \infty.$$

Then the set $\mathfrak{M}(\psi; n)$ of $t \in \mathbb{R}$ for which the inequality

$$|P(t)| < \psi(h(P)) \tag{2.8}$$

holds for infinitely many integral polynomials P of degree n is null.

The restriction that the degree of P be precisely n instead of at most n corresponds to the integer vector $\mathbf{q}$ in (2.4) having non-zero n-th component q_n. It is readily deduced that the theorem also holds for degree at most n, *i.e.*, for $\mathfrak{M}(\psi)$, whence Baker's conjecture is equivalent to $\mathfrak{M}(\psi)$ being null when $\sum_q q^{n-1}\psi(q)$ converges. The theorem was established in [36] and an expanded proof is now given. The approximation function $q^{-n+1}\psi(q)$ in [36] has been changed to $\psi(q)$ and the volume sum altered accordingly to reveal more clearly the Groshev type character of the result. The case $n = 1$ is simply Khintchine's theorem and $n = 2$ corresponds to the parabola being a Groshev type manifold for convergence. The proof makes repeated use of Cantelli's lemma (see §1.2.3).

We begin with some notation and simplifying results. Let ε be a given sufficiently small positive number. Unless otherwise stated, in this section the implied constant in the Vinogradov notation $\ll$ will depend only on the degree n and ε. Since $\sum_q (1/q)$ diverges and the sum in the hypothesis converges, we can assume without loss of generality that $\psi(q) \leqslant cq^{-n}$ for some $c > 0$, so that the inequality

$$|P(t)| < c\,h(P)^{-n} \tag{2.9}$$

is weaker than (2.8). It turns out that only special polynomials of degree n need to be considered. This simplification is on very much the same lines as that in [10] and [208] and so only a brief account will be given.

2.4.1. Reduction to primitive irreducible leading polynomials. If P is not primitive, the coefficients have a highest common factor $a > 1$, so that $P = a\,P_1$ where P_1 is a primitive integer polynomial. Thus $|P_1(t)| < |P(t)|$ and so if there are infinitely many polynomials satisfying (2.8), there are infinitely many primitive polynomials satisfying (2.8). The reduction to irreducible polynomials is not so straightforward and relies on the fact that the height of a product of polynomials is comparable with the product of the heights of the factors (see Lemma 2 of [10]).

LEMMA 2.2. *Let $P_1, \ldots, P_r$ be polynomials and let $P = P_1 \ldots P_r$. Then*

$$h(P) \asymp h(P_1) \ldots h(P_r).$$

The next result is similar to Lemma 8 in [10] and Chapter 1, §5 in [208].

LEMMA 2.3. *Let δ be any positive number. The set E of t such that*

$$|P(t)| < h(P)^{-(n-1)-\delta} \tag{2.10}$$

holds for infinitely many reducible integral polynomials P of degree n is null.

PROOF. Suppose that on the contrary $E = E(\delta)$ has positive measure. Since the set of algebraic numbers is null, there is no loss of generality in taking E to consist of transcendental numbers. Any reducible integer polynomial P of degree n can be expressed as a product $P = P_1 P_2$, where P_1, P_2 are integer polynomials of degree at most $n-1$. Then for each $t \in E$, there exist infinitely many P_1, P_2 such that

$$|P_1(t)P_2(t)| < h(P)^{-n+1-\delta} < c\,h(P_1)^{-n+1-\delta}h(P_2)^{-n+1-\delta} \tag{2.11}$$

for some $c > 0$ by Lemma 2.2. By Sprindžuk's theorem, for any $\delta' > 0$, each inequality

$$|P_1(t)| \geqslant h(P_1)^{-n+1-\delta'} \text{ and } |P_2(t)| \geqslant h(P_2)^{-n+1-\delta'}$$

holds for all but finitely many P_1 and P_2 for almost all points t in E. Since $P_1(t)P_2(t) \neq 0$ for any non-trivial P_1, P_2, there exists a constant $C = C(t) > 0$ such that $|P_1(t)| \geqslant Ch(P_1)^{-n+1-\delta'}$ and $|P_2(t)| \geqslant Ch(P_2)^{-n+1-\delta'}$ hold for all P_1 and P_2. Take $\delta' < \delta$. By the preceding inequalities and (2.11),

$$cC^{-2} > (h(P_1)\,h(P_2))^{\delta-\delta'}$$

for infinitely many P_1 and P_2, which is impossible. □

COROLLARY 2.4. *Let $\psi\colon \mathbb{N} \to \mathbb{R}^+$ be decreasing with the sum $\sum_q q^{n-1}\psi(q)$ convergent. Then the set of t such that the inequality* (2.8) *holds for infinitely many reducible integral polynomials P of degree n is null.*

PROOF. Since $\sum_q q^{n-1}\psi(q)$ converges and ψ is monotonic, we can assume that $\psi(q) \ll q^{-n}$ without loss of generality. Hence the set of t for which (2.8) holds for infinitely many P is a subset of the set E of t for which (2.10) holds with $0 < \delta < 1$ for infinitely many P. But this latter set is null, whence the former set is null. □

A polynomial P is called *leading* [15] if the coefficient of the term of highest power is equal to the height $h(P)$ of P, *i.e.*, if

$$P(x) = a_n x^n + \cdots + a_1 x + a_0 = h(P)x^n + \cdots + a_1 x + a_0.$$

The class of leading, primitive, irreducible integral polynomials of degree n will be denoted by with $a_n = N$ by $\mathfrak{I}_n(N)$ the set of real t such that (2.8) holds for infinitely many polynomials P in $\mathfrak{I}_n$ by $\mathcal{M}(\psi)$. Plainly

$$\mathcal{M}(\psi) \subset \mathfrak{M}(\psi; n)$$

(recall $\mathfrak{M}(\psi;n) = \{t \in \mathfrak{M}(\psi)\colon \deg P = n\}$). But by modifying Lemma 5 in [10] appropriately and following the argument of Lemma 8 in [10], it can be seen that if $|\mathfrak{M}(\psi;n)| > 0$, then $\mathcal{M}(\psi)$ also has positive measure (see [10] or [208, Chapter 1]).

Thus to prove $\mathfrak{M}(\psi; n)$ null, it suffices to prove that $\mathcal{M}(\psi)$ is null and there is no loss of generality in restricting ourselves further to polynomials $P \in \mathfrak{I}_n$.

2.4.2. Root systems and classes. It is more convenient to construct covers using the roots of polynomials. Let $P(x) = a_n x^n + \cdots + a_1 x + a_0$, $a_n \neq 0$, be irreducible. Then any root α of P satisfies

$$\begin{aligned}|\alpha|^n &= |(a_{n-1}\alpha^{n-1} + \cdots + a_1\alpha + a_0)/a_n| \\ &\leqslant (|a_{n-1}||\alpha|^{n-1} + \cdots + |a_1||\alpha| + |a_0|)/|a_n| \\ &\leqslant h(P)(|\alpha|^{n-1} + \cdots + |\alpha| + 1)/|a_n| \\ &\leqslant \frac{h(P)(|\alpha|^n - 1)}{|a_n|(|\alpha| - 1)}\end{aligned}$$

for $|\alpha| \neq 1$. On rearranging, we get that

$$|\alpha| < 1 + \frac{h(P)}{|a_n|} - \frac{h(P)}{|a_n||\alpha|^n}.$$

Thus when $P \in \mathfrak{I}_n$, each root α satisfies $0 < |\alpha| < 2$ ($\alpha \neq 0$ since P is reducible).

Let $\alpha_1, \ldots, \alpha_n$ be the roots of a real polynomial P and define the set $\widetilde{\mathcal{T}}(\alpha_i)$ by

$$\widetilde{\mathcal{T}}(\alpha_i) = \widetilde{\mathcal{T}}(\alpha_i(P)) = \{t \in \mathbb{C}\colon \min\{|t - \alpha_j|\colon j = 1, \ldots, n\} = |t - \alpha_i|\},$$

where $\mathbb{C}$ is the set of complex numbers, *i.e.*, $\widetilde{\mathcal{T}}(\alpha_i)$ is the set of complex numbers with distance from α_i at most the distance from the other zeros. Clearly the sets $\widetilde{\mathcal{T}}(\alpha_i)$, $i = 1, \ldots, n$, are closed with disjoint interiors and with boundaries consisting of straight line segments. The intersection $\mathcal{T}(\alpha_i)$ say of $\widetilde{\mathcal{T}}(\alpha_i)$ with $\mathbb{R}$ is given by

$$\mathcal{T}(\alpha_i) = \widetilde{\mathcal{T}}(\alpha_i) \cap \mathbb{R} = \{t \in \mathbb{R}\colon \min_{1 \leqslant j \leqslant n} |t - \alpha_j| = |t - \alpha_i|\},$$

and is either an interval or empty. Evidently $\cup_i \mathcal{T}(\alpha_i) = \mathbb{R}$, and either there is just one non-empty $\mathcal{T}(\alpha)$ or two of the $\mathcal{T}(\alpha_i)$ are semi-infinite ($n \geqslant 3$). If the root α_i of P is real, then $\alpha_i \in \mathcal{T}(\alpha_i)$ and $\mathcal{T}(\alpha_i)$ is an interval but for a non-real root α_j say, $\mathcal{T}(\alpha_j)$ can be empty. This can happen if there are two real roots near the origin and another which is purely imaginary and of large modulus.

Each polynomial $P \in \mathfrak{I}_n$ satisfying (2.8) has a root α (in the open punctured disc of radius 2) with $t \in \mathcal{T}(\alpha)$. Thus instead of the set of points $t \in \mathbb{R}$ satisfying (2.8) for infinitely many polynomials $P \in \mathfrak{I}_n$, one can consider the set of points t for which there are infinitely many algebraic numbers α satisfying the system

$$\left.\begin{aligned}&\alpha \in P_\alpha^{-1}(0), \\ &t \in \mathcal{T}(\alpha), \\ &|P_\alpha(t)| < \psi(h(P_\alpha)).\end{aligned}\right\} \tag{2.12}$$

The inequality (2.8) can lead, for different roots α, to different systems (2.12). But if (2.8) holds for infinitely many polynomials $P \in \mathfrak{I}_n$, then (2.12) holds for infinitely many pairs (P_α, α) of polynomials in $\mathfrak{I}_n$ and algebraic numbers; and the converse also holds. Thus the set $\mathcal{M}(\psi)$ consists of those t for which there exist infinitely many α with minimal polynomial $P_\alpha \in \mathfrak{I}_n$ and with $|P_\alpha(t)| < \psi(h(P_\alpha))$. In other words, any $t \in \mathcal{M}(\psi)$ lies in $\sigma(P_\alpha; \psi(h(P_\alpha)))$ for infinitely many α, where

$$\sigma(P; \delta) = \{t \in I \colon |P(t)| < \delta\}.$$

Choose $\varepsilon \in (0, 1/2)$. The algebraic numbers α will be partitioned into ε-classes. For later use, we set $\varepsilon_1 = c(n)\varepsilon$, where $c(n)$ is sufficiently small, and write

$$T = [\varepsilon_1^{-1}] + 1 > \varepsilon_1^{-1}.$$

Let α_1 be any root of a polynomial $P \in \mathfrak{I}_n$. The other (distinct) roots $\alpha_2, \ldots, \alpha_n$ can be chosen by changing indices if necessary to *closest distance ordering* given by

$$|\alpha_1 - \alpha_2| \leqslant |\alpha_1 - \alpha_3| \leqslant \cdots \leqslant |\alpha_1 - \alpha_n|. \tag{2.13}$$

For each $j = 2, \ldots, n$, define real numbers μ_j and integers r_j from the relations

$$|\alpha_1 - \alpha_j| = h(P)^{-\mu_j}, \quad (r_j - 1)/T \leqslant \mu_j < r_j/T, \tag{2.14}$$

i.e., $\mu_j = -(\log|\alpha_1 - \alpha_j|)/\log h(P)$ and $r_j = [\mu_j T] + 1$ for $j = 2, \ldots, n$. It follows that $\mu_2 \geqslant \mu_3 \geqslant \cdots \geqslant \mu_n$ and $r_2 \geqslant \cdots \geqslant r_n$. Note that the $\mu_j = \mu_j(\alpha_1)$ and the $r_j = r_j(\alpha_1)$ depend on α_1. It turns out, however, that the number of distinct r_j is finite (at most $(nT+2)^{n-1}$: see Lemma 2.5 below) and depends only on the degree n and T. Write

$$\lambda_j = \lambda_j(\alpha_1) = \frac{r_{j+1} + \cdots + r_n}{T}, \quad 1 \leqslant j \leqslant n-1.$$

Let α_1 be an algebraic number of degree n which is a root of a polynomial in $\mathfrak{I}_n$ and with conjugates $\alpha_2, \ldots, \alpha_n$ in closest distance ordering. Let

$$\mathbf{r} = \mathbf{r}(\alpha) = (r_2, \ldots, r_n)$$

be the integer vector determined by (2.14). Then $\mathbf{r}$ describes the distribution of the conjugates of α_1 in terms of thin annuli centred at α_1 in the disc $\{z \in \mathbb{C} \colon |z| < 2\}$ and so will be called a *root distribution vector*. For each such $\mathbf{r}$, call the set of algebraic numbers with conjugates satisfying (2.14) the *root class* and denote it by $\mathcal{K}(\mathbf{r})$.

LEMMA 2.5. *The number of distinct root distribution vectors* $\mathbf{r}$ *is finite and depends only on* n *and* T *(or* ε*).*

PROOF. Let α_i and α_j be any two roots of the polynomial $P \in \mathfrak{I}_n$. Then $|\alpha_i| < 2$ and so $|\alpha_i - \alpha_j| < 4$. Hence $\mu_i > -\log 4/\log h(P)$ so that $r_i \geqslant 0$ for $h(P)$ sufficiently large. Now we obtain an upper bound for each r_i. Since P is irreducible, its discriminant $\Delta(P)$ satisfies $|\Delta(P)| \geqslant 1$. On the other hand,

$$|\Delta(P)| = h(P)^{2n-2} \prod_{1 \leqslant i < j \leqslant n} |\alpha_i - \alpha_j|^2 \ll h(P)^{2n-2} |\alpha_l - \alpha_k|^2$$

for any distinct l, k, $1 \leqslant l, k \leqslant n$, which implies that

$$1 \leqslant |\Delta(P)|^{1/2} \ll h(P)^{n-1}|\alpha_l - \alpha_k|.$$

Hence if $|\alpha_l - \alpha_k| < h(P)^{-n}$, then for sufficiently large $h(P)$ we get a contradiction. Therefore $|\alpha_1 - \alpha_i| \geqslant h(P)^{-n}$ and so $h(P)^{-\mu_i} \geqslant h(P)^{-n}$, whence

$$r_i/T \leqslant \mu_i + 1/T \leqslant n + 1/T,$$

or $r_i \leqslant nT + 1$. Hence the number of distinct vectors $\mathbf{r}$ is at most $(nT+2)^{n-1}$. □

Following Sprindžuk [210], the class $\mathcal{K}(\mathbf{r})$ is said to be of the *first* or *second* kind according as the root distribution vector $\mathbf{r}$ satisfies

$$r_2/T \leqslant (n - \lambda_2)/2 \text{ or } r_2/T > (n - \lambda_2)/2 \tag{2.15}$$

respectively (the notation differs slightly from [210]). Now given a root distribution vector $\mathbf{r}$, we pass from the system (2.12) and consider the system

$$\left.\begin{array}{l} \alpha \in P_\alpha^{-1}(0), \\ \alpha \in \mathcal{K}(\mathbf{r}), \quad P_\alpha \in \mathfrak{I}_n, \quad t \in \mathcal{T}(\alpha), \\ |P_\alpha(t)| < \psi(h(P_\alpha)) \end{array}\right\} \tag{2.16}$$

instead, considering the algebraic numbers α of degree n as solutions.

For each root distribution vector $\mathbf{r}$, let $\mathcal{M}(\psi; \mathbf{r}) = \mathcal{M}^{(n)}(\psi; \mathbf{r})$ be the set of those $t \in \mathbb{R}$ which satisfy the system (2.16) for infinitely many $\alpha \in \mathcal{K}(\mathbf{r})$ (from now on the exponent n will only be used when needed for clarity). Then

$$\mathcal{M}(\psi) = \bigcup_{\mathbf{r}} \mathcal{M}(\psi; \mathbf{r}),$$

where the union is over the finite number of root distribution vectors $\mathbf{r}$.

2.4.3. Auxiliary estimates. A number of lemmas for later use are now given, starting with some useful bounds for derivatives.

LEMMA 2.6. *Let P be a polynomial in $\mathfrak{I}_n$ and α_1 be any root of P. Then*

$$h(P)^{1-\lambda_1} < |P'(\alpha_1)| \leqslant h(P)^{1-\lambda_1+(n-1)/T}. \tag{2.17}$$

PROOF. For any polynomial $P \in \mathfrak{I}_n$ with roots $\alpha_1, \ldots, \alpha_n$,

$$P'(\alpha_1) = a_n \prod_{i=2}^{n} (\alpha_1 - \alpha_i) = h(P) \prod_{i=2}^{n} (\alpha_1 - \alpha_i).$$

When the roots are ordered as above,

$$|P'(\alpha_1)| = h(P)^{1 - \sum_i \mu_i},$$

so that $h(P)^{1-\lambda_1} < |P'(\alpha_1)| \leqslant h(P)^{1-\lambda_1+(n-1)/T}$, as claimed (see [208, p. 56]). □

The next result is a slight sharpening and extension of Lemma 3 in [10].

LEMMA 2.7. *Let $P \in \mathfrak{I}_n$ and $t \in \mathcal{T}(\alpha_1)$. Then*

$$|t-\alpha_1| \leqslant 2^{n-1}\frac{|P(t)|}{|P'(\alpha_1)|}, \tag{2.18}$$

and

$$|t-\alpha_1| \leqslant \left(\min_{2\leqslant j\leqslant n}\{2^{n-j}\frac{|P(t)|}{|P'(\alpha_1)|}|\alpha_1-\alpha_2|\dots|\alpha_1-\alpha_j|\}\right)^{1/j}, \tag{2.19}$$

where $\alpha_1, \alpha_2, \dots, \alpha_n$ are ordered as before. In particular, when $j = n$,

$$|t-\alpha_1| \leqslant \left(\frac{|P(t)|}{h(P)}\right)^{1/n}.$$

PROOF. By hypothesis, $h(P) = a_n$. The first part of the lemma follows from the identities

$$|t-\alpha_1| = \frac{|P(t)|}{h(P)|t-\alpha_2|\dots|t-\alpha_n|},$$

$$|P'(\alpha_1)| = h(P)|\alpha_1-\alpha_2|\dots|\alpha_1-\alpha_n|$$

and the inequalities

$$|\alpha_1-\alpha_j| \leqslant |t-\alpha_1|+|t-\alpha_j| \leqslant 2|t-\alpha_j|, \quad j=2,\dots,n,$$

which come from the definition of $\mathcal{T}(\alpha_1)$.

Similarly

$$|t-\alpha_1|^j \leqslant |t-\alpha_1|\dots|t-\alpha_j| = \frac{|P(t)|}{h(P)|t-\alpha_{j+1}|\dots|t-\alpha_n|} \tag{2.20}$$

$$\leqslant 2^{n-j}\frac{|P(t)|}{h(P)|\alpha_1-\alpha_{j+1}|\dots|\alpha_1-\alpha_n|} \tag{2.21}$$

$$= 2^{n-j}\frac{|P(t)|}{|P'(\alpha_1)|}|\alpha_1-\alpha_2|\dots|\alpha_1-\alpha_j|, \tag{2.22}$$

from which (2.19) follows. □

By taking $j = n$ it can be seen that for each $t \in \mathbb{R}$ satisfying (2.9),

$$|t| \leqslant |t-\alpha_1|+|\alpha_1| < |P(t)|^{1/n}+2 < 3$$

for $h(P)$ sufficiently large, whence $t \in (-3,3)$. This means that for each $\mathbf{r}$,

$$\mathcal{M}(\psi;\mathbf{r}) \subset \mathcal{M}(\psi) \subset (-3,3).$$

LEMMA 2.8. *Let $P \in \mathfrak{I}_n$. Then for each $j = 1, \dots, n-1$,*

$$|P^{(j)}(\alpha_1)| \ll h(P)^{1-\lambda_j+(n-j)\varepsilon_1}.$$

PROOF. The proof is in [32]. The idea is that when $n \geqslant 2$, the root α_1 of the polynomial P satisfies

$$P'(\alpha_1) = h(P)\,(\alpha_1 - \alpha_2)\dots(\alpha_1 - \alpha_n)$$

and when $n \geqslant 3$,

$$P''(\alpha_1) = 2h(P)\,(\alpha_1 - \alpha_3)\dots(\alpha_1 - \alpha_n) + \dots + 2h(P)\,(\alpha_1 - \alpha_2)\dots(\alpha_1 - \alpha_{n-1}),$$

so that

$$|P''(\alpha_1)| = 2\,h(P)^{1-\sum_{i=3}^{n}\mu_i} + \dots + 2\,h(P)^{1-\sum_{i=2}^{n-1}\mu_i} \ll h(P)^{1-\lambda_2+(n-2)/T}.$$

□

The next two results involve the modulus of polynomials.

LEMMA 2.9. *Let P, Q be two integral polynomials of degree at most n without common roots. Let δ be a positive real number and let $N \geqslant N(\delta, n)$ a sufficiently large real number. Suppose*

$$\max\{h(P), h(Q)\} \leqslant N^{\mu}$$

and let $I \subset (-n, n)$ be an interval of length $|I| = N^{-\eta}$, $\eta > 0$. If there exists a $v > 0$ such that for all x in I,

$$\max\{|P(x)|, |Q(x)|\} < N^{-v}, \qquad (2.23)$$

then $v + \mu + 2\max(v + \mu - \eta,\, 0) < 2n\mu + \delta$.

This is a consequence of the fact that two coprime polynomials have no roots in common (for the proof see Lemma 12 in [33]).

LEMMA 2.10. *Let I be an interval in $\mathbb{R}$ and let B be a measurable subset of I with $|B| \geqslant |I|/k$, where $k \in \mathbb{N}$. Suppose that*

$$|P(t)| < h(P)^{-v}$$

for all $t \in B$, where $v > 0$ and $\deg P \leqslant n$. Then for all $t \in I$,

$$|P\,(t)| < (3k)^n (n+1)^{(n+1)} h(P)^{-v}.$$

PROOF. Divide the interval I into $3k(n+1)$ intervals $I_1, \dots, I_{3k(n+1)}$ of equal length. Since $|B| \geqslant |I|/k$, B intersects at least $3n+3$ intervals. We denote these intervals by $I_{s_1}, \dots, I_{s_l}$, $l \geqslant 3n+3$. Consider three consecutive intervals $I_{s_j}, I_{s_{j+1}}, I_{s_{j+2}}$ and choose $t \in B \cap I_{s_{j+1}}$. In this way we can choose $n+1$ points $t_1, \dots, t_{n+1}$ that satisfy the conditions

$$|t_i - t_j| \geqslant \frac{1}{3k(n+1)}|I|, \; i \neq j,$$

and at which $|P\,(t_i)| < h(P)^{-v}$. By the Lagrange interpolation formula,

$$P\,(t) = \sum_{l=1}^{n+1} \frac{P(t_l)(t-t_1)\dots(t-t_{l-1})(t-t_{l+1})\dots(t-t_{n+1})}{(t_l-t_1)\dots(t_l-t_{l-1})(t_l-t_{l+1})\dots(t_l-t_{n+1})}.$$

Since $|t-t_l| \leqslant |I|$ and $|P(t_l)| < h(P)^{-v}$ for all $t \in I$, we obtain from the representation of $P(t)$ that

$$|P(t)| < \sum_{l=1}^{n+1} h(P)^{-v} \frac{|I|^n}{\big(3k(n+1)\big)^{-n}|I|^n} = (3k)^n(n+1)^{n+1}h(P)^{-v}.$$

□

The final result is Lemma 7 of [10].

LEMMA 2.11. *For each positive integer H, denote by $\mathcal{U}(H)$ a finite set of real closed intervals. Let $\mathcal{V}(H)$ denote a subset of $\mathcal{U}(H)$ such that for each $I \in \mathcal{V}(H)$ there exists a $J \neq I, J \in \mathcal{U}(H)$, for which $|I \cap J| \geqslant |I|/2$. Let $V(H)$ denote the union of the points of the intervals I of $\mathcal{V}(H)$ and let $v(H)$ denote the union of the intervals $I \cap J$. Further, let W and w denote the set of points contained in infinitely many $V(H)$ and in infinitely many $v(H)$ respectively. Then if w is null so is W.*

Theorem 2.1 is proved by using the sets $\mathcal{M}(\psi; \mathbf{r})$ of points $t \in (-3,3)$ such that the system (2.16) holds for infinitely many $\alpha \in \mathcal{K}(\mathbf{r})$ for various ranges of the coefficients in the root distribution vector $\mathbf{r}$. In order to illustrate some of the ideas involved, we consider the simple example of the parabola (in fact smooth planar curves with non-zero curvature almost everywhere are of Groshev type for convergence [37]).

2.4.4. The parabola again. We show that the parabola is of Groshev type for convergence. By the above discussion, it suffices to prove that if $\sum_q q\psi(q)$ converges then the set $\mathcal{M}^{(2)}(\psi)$ of points $t \in (-3,3)$ such that $|Q(t)| < \psi(h(Q))$ holds for infinitely many integer quadratics $Q(x) = h(Q)x^2 + a_1x + a_0$ in $\mathfrak{I}_2$ is null.

Let $\alpha(Q)$ and $\alpha'(Q)$ be the two roots of the quadratic Q. The proof uses a cover based on neighbourhoods of these roots. Define

$$\mathcal{T}(\alpha(Q)) = \{t \in \mathbb{R} \colon |t-\alpha'(Q)| \geqslant |t-\alpha(Q)|\}.$$

When Q has real roots,

$$\alpha(Q) \in \mathcal{T}(\alpha(Q)), \alpha'(Q) \in \mathcal{T}(\alpha'(Q))$$

and $\mathcal{T}(\alpha(Q)) = \mathcal{T}(\alpha'(Q)) = \mathbb{R}$ otherwise. Suppose $t \in \mathcal{M}^{(2)}(\psi)$ and that t lies in $\mathcal{T}(\alpha(Q))$ for some Q in $\mathfrak{I}_2(N)$, the set of quadratics in $\mathfrak{I}_2$ with leading coefficient N. Then by Lemma 2.7

$$|t-\alpha(Q)| = \frac{|Q(t)|}{h(Q)|t-\alpha'(Q)|} \leqslant \frac{2|Q(t)|}{h(Q)|\alpha(Q)-\alpha'(Q)|} < \frac{2\psi(h(Q))}{\sqrt{\Delta(Q)}} = r_Q,$$

where $\Delta(Q)$ is the discriminant of Q. Thus when $\alpha(Q)$ is real,

$$t \in (\alpha(Q)-r_Q, \alpha(Q)+r_Q) = I_Q$$

say, an interval of length $2r_Q$. Similarly if $t \in \mathcal{T}(\alpha'(Q))$ for $Q \in \mathfrak{I}_2(N)$, then $t \in (\alpha'(Q) - r_Q, \alpha'(Q) + r_Q) = I'_Q$. Thus t lies in J_Q, where J_Q is a union of at most two intervals each of length $2r_Q$ when α is real and is an interval of length at most $2r_Q$ otherwise. Hence

$$\mathcal{M}^{(2)}(\psi) \subseteq \{t \in (-3,3) \colon t \in J_Q \text{ for infinitely many } \ Q \in \mathfrak{I}_2\}.$$

But

$$\begin{aligned}\sum_{Q\in\mathfrak{I}_2} |J_Q| &\ll \sum_{N=1}^{\infty} \sum_{Q\in\mathfrak{I}_2(N)} r_Q \ll \sum_{N=1}^{\infty} \sum_{Q\in\mathfrak{I}_2(N)} \psi(N)\Delta(Q)^{-1/2}\\ &\ll \sum_{N=1}^{\infty} \psi(N) \sum_{|a_1|\leqslant N} \sum_{\substack{|a_0|\leqslant N\\ a_0\neq a_1^2/4N}} |a_1^2 - 4a_0N|^{-1/2}.\end{aligned}$$

Now for each fixed a_1,

$$\sum_{a_0=-N}^{N} |a_1^2 - 4a_0N|^{-1/2} \ll 1 + \sum_{k=1}^{N} (kN)^{-1/2} \ll N^{-1/2}N^{1-1/2} \ll 1,$$

whence

$$\sum_{N=1}^{\infty} \sum_{Q\in\mathfrak{I}_2(N)} |J_Q| \ll \sum_{N=1}^{\infty} \psi(N) \sum_{|a_1|\leqslant N} 1 \ll \sum_{N=1}^{\infty} N\psi(N).$$

Thus if the sum $\sum_q q\psi(q)$ converges, then by Cantelli's lemma, $\mathcal{M}^{(2)}(\psi)$ is null and so $\mathfrak{M}^{(2)}(\psi)$ is null.

2.4.5. The proof of Baker's conjecture. We gather together some definitions which will be used. Let

(1) $\mathfrak{I}_n(N) = \{P \in \mathfrak{I}_n \colon h(P) = N\}$;
(2) $\mathfrak{I}_n(N\,;\mathbf{r})$ be those $P \in \mathfrak{I}_n(N)$ with root distribution vector $\mathbf{r}$;
(3) $E(\mathbf{r}) = \mathcal{M}(f_n;\mathbf{r})$ where $f_n(q) = q^{-n}$, thus if $\psi(q) \leqslant q^{-n}$ for each sufficiently large q, then $\mathcal{M}(\psi;\mathbf{r}) \subseteq E(\mathbf{r})$.

We make a further definition: let J be an interval; a polynomial P *belongs to* J if there exists a $t \in J$ such that (2.9) holds. Recall that $n \geqslant 3$ and that the exponent n is omitted from $\mathfrak{M}^{(n)}(\psi)$ unless there is any ambiguity.

PROPOSITION 2.12. *Let*

$$\mathcal{M}_1(\psi) = \bigcup_{\mathbf{r}} \mathcal{M}(\psi;\mathbf{r}),$$

where the union is over root distribution vectors $\mathbf{r} = (r_2,\ldots,r_n)$ *satisfying*

$$n - 1 + 2n\varepsilon_1 < r_2T^{-1} + \lambda_1.$$

Then $\mathcal{M}_1(\psi)$ *is null.*

PROOF. From inequality (2.8) we pass to the weaker inequality (2.9) and consider the set $E(\mathbf{r})$ (see (3) in §2.4.5). The inequality $r_2T^{-1}+\lambda_1 \geqslant n$ corresponds to root classes $\mathcal{K}(\mathbf{r})$ of the second kind (2.15). This case is covered in [208, Chapter 3, §3]) so we assume that

$$n-1+2n\varepsilon_1 < r_2T^{-1}+\lambda_1 < n. \tag{2.24}$$

Let ℓ be a non-negative integer. Divide the interval Ω into $6[2^{\ell\theta}]$ subintervals J, having equal length $|J| = 1/[2^{\ell\theta}] \asymp 2^{-\ell\theta}$ where

$$\theta = n+1-\lambda_1-\varepsilon_1/2.$$

For each $\ell = 0,1,2,\ldots$, define

$$\mathfrak{I}_n^{(\ell)} = \{P \in \mathfrak{I}_n \colon 2^\ell \leqslant h(P) < 2^{\ell+1}\} = \bigcup_{2^\ell \leqslant N < 2^{\ell+1}} \mathfrak{I}_n(N).$$

Two non-conjugate algebraic numbers α_1, β_1 in $\mathcal{K}(\mathbf{r})$ cannot satisfy polynomials P, Q in $\mathfrak{I}_n^{(\ell)}$ belonging to the same interval J. For suppose to the contrary that there exists an interval J to which the polynomials P, $Q \in \mathfrak{I}_n^{(\ell)}$ belong. Then there exist points $\omega_1, \omega_2 \in J$, such that

$$\max\{|P(\omega_1)|,\ |Q(\omega_2)|\} \ll 2^{-\ell n}. \tag{2.25}$$

It follows from Lemmas 2.7 and 2.6 and (2.25) that

$$\max\left\{|\omega_1-\alpha_1|,\ |\omega_2-\beta_1|\right\} \ll 2^{\ell(-n-1+\lambda_1)}. \tag{2.26}$$

Since $|\omega_1-\omega_2| < |J| \ll 2^{-\ell\theta}$,

$$\begin{aligned} |\alpha_1-\beta_1| &\leqslant |\alpha_1-\omega_1| + |\omega_1-\omega_2| + |\omega_2-\beta_1| \\ &\ll 2^{\ell(-n-1+\lambda_1)} + 2^{-\ell\theta} \ll 2^{-\ell\theta} \end{aligned}$$

and by (2.14) for $j \geqslant 2$,

$$|\alpha_1-\beta_j| < |\alpha_1-\beta_1| + |\beta_1-\beta_j| \ll 2^{-\ell\theta} + 2^{\ell(-r_jT^{-1}+\varepsilon_1)},$$

whence by the definition of θ and (2.24) (which imply that $r_2/T < n-\lambda_1$),

$$|\alpha_1-\beta_j| \ll 2^{\ell(-r_jT^{-1}+\varepsilon_1)}.$$

It follows that

$$\prod_{j=1}^{n} |\alpha_1-\beta_j| \ll 2^{\ell(-\theta-\lambda_1+(n-1)\varepsilon_1)} \tag{2.27}$$

and similarly

$$\begin{aligned} \prod_{j=1}^{n} |\alpha_2-\beta_j| &\ll \prod_{j=1}^{n}(|\alpha_2-\alpha_1| + |\alpha_1-\beta_1| + |\beta_1-\beta_j|) \qquad (2.28) \\ &\ll 2^{\ell(-r_2T^{-1}-\lambda_1+n\varepsilon_1)}. \qquad (2.29) \end{aligned}$$

It turns out that small estimates for $|\alpha_i - \beta_j|$ are needed only when $i = 1, 2$; for $i \geqslant 3$, $j = 1, \ldots, n$, the crude estimate $|\alpha_i - \beta_j| \ll 1$ suffices.

Since the integral polynomials P, $Q \in \mathfrak{I}_n^{(\ell)}$ are irreducible, they do not have common roots and so their resultant $R(P, Q)$ satisfies

$$|R(P, Q)| = h(P)^n h(Q)^n \prod_{i=1}^{n} \prod_{j=1}^{n} |\alpha_i(P) - \alpha_j(Q)| \geqslant 1.$$

The above inequalities (2.27) and (2.29) imply

$$1 \leqslant |R(P, Q)| \ll 2^{\ell(2n-\theta-\lambda_1+(n-1)\varepsilon_1-r_2T^{-1}-\lambda_1+n\varepsilon_1)+2n} \ll 2^{-\ell\varepsilon_1/2},$$

which is impossible for large ℓ.

Thus for each interval J, there is at most one algebraic number $\alpha_J \in \mathcal{K}(\mathbf{r})$, where $\mathbf{r}$ satisfies (2.24), satisfying the polynomial $P_J \in \mathfrak{I}_n^{(\ell)}$ belonging to J. Relabel α_J as α_1 and the other roots according to the closest distance ordering (2.13). Let

$$\sigma(\alpha_1) = \{t \in \mathcal{T}(\alpha_1) \cap (-3, 3) \colon |P_{\alpha_1}(x)| < h(P_{\alpha_1})^{-n}\},$$

where $P_{\alpha_1} \in \mathfrak{I}_n^{(\ell)}$ is satisfied by α_1. Then by Lemmas 2.7 and 2.6,

$$|\sigma(\alpha_1)| \ll 2^{-\ell(n+1-\lambda_1)}.$$

Let $E^{(\ell)}(\mathbf{r})$ be the set of $t \in (-3, 3)$ such that there exists an algebraic number $\alpha \in \mathcal{K}(\mathbf{r})$ satisfying the polynomial $P \in \mathfrak{I}_n^{(\ell)}$ and satisfying (2.9), *i.e.*,

$$|P(t)| < h(P)^{-n},$$

where $\mathbf{r}$ is in the range (2.24). Then

$$E(\mathbf{r}) = \{t \in (-3, 3) \colon t \in E^{(\ell)}(\mathbf{r}) \text{ for infinitely many } \ell\}.$$

Now for each $\ell = 0, 1, 2, \ldots$, the subset $E^{(\ell)}(\mathbf{r})$ is covered by at most $\ll 2^{\ell\theta}$ (the number of intervals J) intervals $\sigma(\alpha_1)$, each of length at most $2^{-\ell(n+1-\lambda_1)}$. Hence

$$|E^{(\ell)}(\mathbf{r})| \ll 2^{-\ell(n+1-\lambda_1-\theta)} \ll 2^{-\ell(\varepsilon_1/2)}$$

and so

$$\sum_{\ell=0}^{\infty} |E^{(\ell)}(\mathbf{r})| \ll \sum_{\ell=0}^{\infty} 2^{-\ell(n+1-\lambda_1-\theta)} \ll \sum_{\ell=0}^{\infty} 2^{-\ell(\varepsilon_1/2)} < \infty.$$

By the Cantelli's lemma, $E(\mathbf{r})$ is null and since there are only finitely many root distribution vectors $\mathbf{r}$,

$$|\mathcal{M}_1(\psi)| \leqslant \sum_{\mathbf{r}} |\mathcal{M}(\psi; \mathbf{r})| \leqslant \sum_{\mathbf{r}} |E(\mathbf{r})| = 0,$$

as claimed. □

PROPOSITION 2.13. *Let*

$$\mathcal{M}_2(\psi) = \bigcup_{\mathbf{r}} \mathcal{M}(\psi; \mathbf{r}),$$

where the union is over root distribution vectors $\mathbf{r} = (r_2, \ldots, r_n)$ *satisfying*

$$2 - \varepsilon/2 \leqslant r_2 T^{-1} + \lambda_1 \leqslant n - 1 + 2n\varepsilon_1. \tag{2.30}$$

Then $\mathcal{M}_2(\psi)$ *is null.*

PROOF. The proof here is broken down into various cases and subcases and we again pass to the set $E(\mathbf{r})$. Put

$$\theta = n + 1 - r_2 T^{-1} - \lambda_1 \tag{2.31}$$

and consider first the case when

$$\{\theta\} > \varepsilon\,(> 0) \tag{2.32}$$

where $\{\theta\}$ is the fractional part of θ. Then the left hand inequality of (2.30) and the definition of θ imply $n - [\theta] > 1$ (and hence $\geqslant 2$). Divide the interval $(-3, 3)$ into $6[N^{\theta_1}]$ intervals J of equal length $[N^{\theta_1}]^{-1}$, where

$$\theta_1 = \frac{r_2}{T} + \frac{4\{\theta\}}{5} + (n+1)\varepsilon_1.$$

Let $\mathcal{N}_J$ be the number of algebraic numbers in $\mathcal{K}(\mathbf{r})$ which satisfy polynomials $P \in \mathfrak{I}_n(N)$ belonging to J.

(a) Suppose that for each $N \geqslant N_0$ and each J, $\mathcal{N}_J \ll N^{\nu}$, where

$$\nu = [\theta] - 1 + \{\theta\}/5 - \varepsilon/10.$$

Let

$$\sigma(\alpha_1) = \{t \in \mathcal{T}(\alpha_1) \cap (-3,3) \colon |P_{\alpha_1}(t)| \ll N^{-n}\}.$$

Then by the definitions of θ, θ_1 and ν, the set $E(N; \mathbf{r})$ of t satisfying (2.9) for $P \in \mathfrak{I}_n(N)$ has Lebesgue measure

$$|E(N; \mathbf{r})| \ll \sum_J |\sigma(\alpha_1)| \ll N^{\theta_1 + \nu - n - 1 + \lambda_1} \ll N^{-1-\varepsilon/20}$$

(we have used $|\sigma(\alpha_1)| \ll N^{-(n+1-\lambda_1)}$ and $\varepsilon > 20(n+1)\varepsilon_1$). But each $t \in E(\mathbf{r})$ falls into $E(\mathbf{r}; N)$ for infinitely many N and

$$\sum_{N=1}^{\infty} |E(\mathbf{r}; N)| \ll \sum_{N=1}^{\infty} N^{-1-\varepsilon/20} < \infty,$$

whence by Cantelli's lemma, $|E(\mathbf{r})| = 0$ and it follows that $\mathcal{M}_2(\psi)$ is null.

(b) On the other hand suppose for infinitely many N, there exists an interval J such that $\mathcal{N}_J \gg N^\nu$. Fix one such interval J and let $P_1, \dots, P_{\mathcal{N}_J} \in \mathfrak{I}_n(N;\mathbf{r})$ belong to J. Two such polynomials

$$P_i(x) = Nx^n + a_{n-1}^i x^{n-1} + \cdots + a_0^i,$$
$$P_j(x) = Nx^n + a_{n-1}^j x^{n-1} + \cdots + a_0^j,$$

where $1 \leqslant i < j \leqslant \mathcal{N}_J$, are included in the same class if

$$a_{n-1}^i = a_{n-1}^j, \dots, a_{n-[\theta]+1}^i = a_{n-[\theta]+1}^j.$$

The number of different classes is $\ll N^{[\theta]-1}$, so by Dirichlet's pigeonhole principle, there are $\ell_J + 1$ polynomials $P_0, \dots, P_{\ell_J}$ say, where $\ell_J \gg N^{\{\theta\}/5-\varepsilon/10}$, amongst the ν_J polynomials belonging to the same class, where $\nu_J \gg N^\nu$. Form the ℓ_J differences

$$R_1(x) = P_1(x) - P_0(x), \dots, R_{\ell_J}(x) = P_{\ell_J}(x) - P_0(x).$$

All these polynomials R_i are distinct, with degree at most $n - [\theta]$ (> 1 whence $\deg R_i \geqslant 2$) and height at most $2N$.

The Taylor series for the polynomial P about the root α_1 is

$$P(\omega) = P'(\alpha_1)(\omega - \alpha_1) + \cdots + \frac{P^{(n)}(\alpha_1)}{n!}(\omega - \alpha_1)^n. \tag{2.33}$$

When P belongs to J, there exists a point $t_0 \in J$ such that $|P(t_0)| \ll N^{-n}$, whence by Lemmas 2.6 and 2.7 we obtain

$$|t_0 - \alpha_1| \ll N^{-n-1+\lambda_1}.$$

For each $\omega \in J$, we have $|\omega - t_0| \leqslant |J|$ and by (2.32)

$$|\omega - \alpha_1| \ll N^{-(r_2/T)-(4\{\theta\}/5)-(n+1)\varepsilon_1}.$$

Using Lemma 2.8 and the last inequality we have

$$\max\{|P^{(j)}(\alpha_1)(\omega-\alpha_1)^j| : j = 1, \dots, n\} \ll N^{-(n-\theta)-4\{\theta\}/5}.$$

The last inequality and (2.33) imply that for $\omega \in J$,

$$|P(\omega)| \ll N^{-(n-\theta)-4\{\theta\}/5}.$$

Since this last inequality is valid for any $P \in \mathfrak{I}_n(N;\mathbf{r})$ which also belongs to J, it follows that each polynomial R_i, $1 \leqslant i \leqslant \ell_J$, satisfies

$$|R_i(\omega)| \ll N^{-(n-\theta)-4\{\theta\}/5}. \tag{2.34}$$

(i) Suppose that for each i, $R_i(\omega) = a_i R(\omega)$, $a_i \in \mathbb{Z}$ (*i.e.,* we suppose the R_i are not primitive). Since the R_i are all different, so are the a_i. Let $\tilde{a}$ have the maximum modulus of the a_i, so that $|\tilde{a}| \geqslant \ell_J/2$. Then the polynomial $S(\omega) = \tilde{a}R(\omega)$ has height $h(S) = |\tilde{a}|h(R) \leqslant 2N$. Hence $h(R) \ll N^{1-\{\theta\}/5+\varepsilon/10}$ and so

$$N^{-1} \ll h(R)^{-1/(1-\{\theta\}/5+\varepsilon/10)}$$

and the inequality (2.34) implies

$$|R(\omega)| \ll h(R)^{-\mu_1}, \tag{2.35}$$

where by (2.34),

$$\mu_1 = \frac{n-\theta+4\{\theta\}/5}{1-\{\theta\}/5+\varepsilon/10}.$$

Since $0<\varepsilon<\{\theta\}$ (or by (2.32)), $\mu_1 > n-[\theta] \geqslant \deg R$. There is at least one such polynomial R for each sufficiently large N. But by Sprindžuk's theorem [208], the set of ω for which there are infinitely many polynomials R satisfying (2.35) is null.

(ii) Next suppose that there exists at least one polynomial S such that for some i, $R_i(t) = S(t)$ with R and S relatively prime. Suppose there is an infinite sequence N_r, $r=1,2,\ldots$, such that for each N_r there exists at least one polynomial R_i which is reducible. The polynomials in the sequence are all different by (2.34). Without loss of generality, $n-\theta+4\{\theta\}/5 > n-[\theta]-1$; Lemma 2.3 applied to the reducible polynomials R_i implies that the set of these points t is null.

(iii) Finally suppose that for infinitely many N, all the polynomials R_i are irreducible. Then among the ℓ_J polynomials no two have common rational roots. Choose any two and use Lemma 2.9 with n replaced by $n-[\theta]$. By the condition (2.32) it is sufficient to assume

$$v = n-\theta+4\{\theta\}/5,$$
$$\eta = r_2T^{-1}+4\{\theta\}/5+(n+1)\varepsilon_1 = \theta_1,\ \mu=1.$$

Hence

$$r_2T^{-1}+3\lambda_1+4\{\theta\}/5-2(n+1)\varepsilon_1 \leqslant 2r_2T^{-1}+2\lambda_1-2+2\{\theta\}+\delta.$$

As $\lambda_1 \geqslant r_2T^{-1}$, the last inequality cannot hold if δ is sufficiently small.

The other case $0 \leqslant \{\theta\} \leqslant \varepsilon$ involves slight changes in the choice of parameters. Note that in this case (2.30) and (2.31) imply that $[\theta] \geqslant 2$ and so $n-[\theta]+1 \leqslant n-1$. Putting $\eta = r_2T^{-1}-4/5+(n+1)\varepsilon_1$ and $\nu = [\theta]-9/5-\varepsilon/10$ and carrying out the corresponding argument completes the proof of the proposition. □

PROPOSITION 2.14. *Let*

$$\mathcal{M}_3(\psi) = \bigcup_{\mathbf{r}} \mathcal{M}(\psi;\mathbf{r}),$$

where the union is over root distribution vectors $\mathbf{r} = (r_2,\ldots,r_n)$ *satisfying*

$$\varepsilon < r_2T^{-1}+\lambda_1 < 2-\frac{\varepsilon}{2}. \tag{2.36}$$

Then $\mathcal{M}_3(\psi)$ *is null.*

PROOF. Let $\mathcal{M}(\mathbf{r};N)$ be the set of points t in $\mathcal{M}(\psi;\mathbf{r})$ such that (2.8) holds for a polynomial $P \in \mathfrak{I}_n(N)$. Then

$$\mathcal{M}(\psi;\mathbf{r}) = \{t \in (-3,3)\colon t \in \mathcal{M}(\mathbf{r};N) \text{ for infinitely many } N \in \mathbb{N}\}.$$

Let $\mathbf{a} = (a_{n-1}, \dots, a_2)$ be an integer vector and write $\mathcal{M}(N; \mathbf{r}; \mathbf{a})$ for the set of $t \in \mathcal{M}(N; \mathbf{r})$ for which there exists an algebraic number α_1 satisfying a polynomial $P \in \mathfrak{I}_n(N)$ of the form

$$P(t) = Nx^n + a_{n-1}x^{n-1} + \cdots + a_2x^2 + a_1x + a_0.$$

The set of such polynomials P will be denoted by $\mathfrak{I}_n(N\,; \mathbf{r}, \mathbf{a})$.

Let α_1 satisfy the polynomial $P \in \mathfrak{I}_n(N\,; \mathbf{r}, \mathbf{a})$, let

$$B(\alpha_1) = \{t \in (-3,3) \colon |t - \alpha_1| < 2^n \psi(N)|P'(\alpha_1)|^{-1}\}, \tag{2.37}$$

and let

$$\widetilde{B}(\alpha_1) = \{t \in (-3,3) \colon |t - \alpha_1| < 2^n |P'(\alpha_1)|^{-1} N^{-1}\}.$$

It is evident that $B\,(\alpha_1) \subset \widetilde{B}\,(\alpha_1)$ and that $|B(\alpha_1)| < N\psi(N)|\widetilde{B}(\alpha_1)|$. The inequality (2.18) implies that each $t \in \mathcal{T}(\alpha_1)$ satisfying the inequalities $|P(t)| < \psi(N)$ and $|P(t)| < N^{-1}$ lies in the sets $B(\alpha_1)$ and $\widetilde{B}(\alpha_1)$ respectively.

Let $t \in \widetilde{B}(\alpha_1)$. It follows immediately from the definition of $\widetilde{B}(\alpha_1)$ that when $t \in \widetilde{B}(\alpha_1)$,

$$|P'(\alpha_1)(t - \alpha_1)| < 2^n N^{-1}.$$

Now for $i \geqslant 2$, $\lambda_1 - \lambda_i = (r_2 + \cdots + r_i)/T \leqslant (i-1)r_2/T$, whence by (2.36),

$$i\lambda_1 - \lambda_i \leqslant (i-1)(r_2 T^{-1} + \lambda_1) < (i-1)(2 - \varepsilon/2).$$

Hence by Lemmas 2.8, 2.6 and 2.7,

$$\max_{2 \leqslant i \leqslant n} |P^{(i)}\,(\alpha_1)(t - \alpha_1)^i| \ll \max_{2 \leqslant i \leqslant n} N^{1-\lambda_i+(n-i)\varepsilon_1} \left(N^{-1+\lambda_1} N^{-1}\right)^i \ll N^{-1-\varepsilon_1},$$

and so by Taylor's expansion (2.33) for P around α_1, $|P(t)| \ll N^{-1}$.

We shall call the interval $\widetilde{B}(\alpha_1)$ *inessential* if there is an algebraic number β_1 with minimal polynomial Q in the class $\mathfrak{I}_n\,(N; \mathbf{r}; \mathbf{a})$ such that

$$|\widetilde{B}(\alpha_1) \cap \widetilde{B}(\beta_1)| \geqslant |\widetilde{B}(\alpha_1)|/2$$

and *essential* otherwise. When the interval $\widetilde{B}(\alpha_1)$ is inessential, then on the interval $I = \widetilde{B}(\alpha_1) \cap \widetilde{B}(\beta_1)$, the first degree polynomial $R\,(t) = P\,(t) - Q\,(t)$ has height $h(R) \geqslant 1$ and satisfies

$$|R\,(t)| \leqslant |P(t)| + |Q(t)| \ll N^{-1}. \tag{2.38}$$

As the length $|I|$ of the interval I satisfies

$$2^n N^{-1}|P'(\alpha_1)|^{-1} \leqslant |I| \leqslant 2^{n+1} N^{-1}|P'(\alpha_1)|^{-1},$$

and since R is not constant, the height $h(R)$ of the polynomial R satisfies

$$1 \leqslant h(R) \ll |P'(\alpha_1)| \ll N^{1-\lambda_1+(n-1)\varepsilon_1}. \tag{2.39}$$

Hence by (2.38) it follows that for all $t \in I$

$$|R\,(t)| \ll h(R)^{-1/(1-\lambda_1+(n-1)\varepsilon_1)} < h(R)^{-1-\eta}$$

for some $\eta > 0$, providing $0 < 1 - \lambda_1 + (n-1)\varepsilon_1 < 1$. Hence by Khintchine's theorem, the set of such points t is null and hence the set of points t which lie in infinitely many inessential intervals $\widetilde{B}(\alpha_1)$ is also null by Lemma 2.11.

Now $\lambda_1 \geqslant r_2/T$ and the inequality $1 - \lambda_1 + (n-1)\varepsilon_1 < 1$ is equivalent to $\lambda_1 > (n-1)\varepsilon_1 = (n-1)/T$. Assume the contrary, *i.e.*, assume $\lambda_1 \leqslant (n-1)/T$. Then

$$\frac{r_2}{T} + \lambda_1 \leqslant 2\lambda_1 \leqslant \frac{2(n-1)}{T}.$$

But by the range (2.36), $\lambda_1 + r_2/T > \varepsilon > 3n/T$, a contradiction.

By (2.39),

$$1 \leqslant |h(R)| \ll |P'(\alpha_1)| \tag{2.40}$$

(whence $|P'(\alpha_1)| \gg 1$) while by Lemma 2.6 $|P'(\alpha_1)| \leqslant N^{1-\lambda_1+(n-1)/T}$, which implies $1 - \lambda_1 + (n-1)/T \geqslant 0$ for N sufficiently large.

Suppose $1 - \lambda_1 + (n-1)\varepsilon_1 = 0$. Then by Lemma 2.6, $|P'(\alpha_1)| = 1$. Hence by (2.40) the number of different linear polynomials R is independent of N, which implies that the number of polynomials in $\mathfrak{I}_n(N, \mathbf{r}, \mathbf{a})$ is also independent of N. Now

$$|B(\alpha_1)| \ll \psi(N)|P'(\alpha_1)|^{-1} \ll N^{-1+\lambda_1}\psi(N) \ll N^{(n-1)/T}\psi(N),$$

whence

$$\begin{aligned}\sum_{P\in\mathfrak{I}_n(N,\mathbf{r})} |B(\alpha_1)| &= \sum_{\mathbf{a}} \sum_{P\in\mathfrak{I}_n(N,\mathbf{r},\mathbf{a})} N^{(n-1)/T}\psi(N)\\ &\ll \sum_{\mathbf{a}} N^{(n-1)/T}\psi(N) \ll N^{n-2+(n-1)/T}\,\psi(N).\end{aligned}$$

Providing T is sufficiently large, the series $\sum_{N=1}^{\infty} N^{n-2+(n-1)/T}\psi(N)$ converges, and so by the Cantelli's lemma implies that the set of t falling into infinitely many $B(\alpha_1)$ is null.

If $\widetilde{B}(\alpha_1)$ is an essential interval, then each point $t \in (-3, 3)$ belongs to at most three essential intervals and hence

$$\sum_{P_{\alpha_1}\in\mathfrak{I}(N,\mathbf{r},\mathbf{a})} |\widetilde{B}(\alpha_1)| \leqslant 19.$$

From the inequality $|B(\alpha_1)| \leqslant N\,\psi(N)\,|\widetilde{B}(\alpha_1)|$, it follows that

$$\sum_{P\in\mathfrak{I}_n(N,\mathbf{r})} |B(\alpha_1)| = \sum_{\mathbf{a}} \sum_{P\in\mathfrak{I}_n(N,\mathbf{r},\mathbf{a})} |B(\alpha_1)| \ll \sum_{\mathbf{a}} N\,\psi(N) \ll N^{n-1}\,\psi(N).$$

Since the series $\sum_{N=1}^{\infty} N^{n-1}\psi(N)$ converges, the Cantelli's lemma implies that the set of t falling into infinitely many essential intervals $\widetilde{B}(\alpha_1)$ is null. $\square$

PROPOSITION 2.15. *Let*

$$\mathcal{M}_3(\psi) = \bigcup_{\mathbf{r}} \mathcal{M}(\psi; \mathbf{r}),$$

where the union is over root distribution vectors $\mathbf{r} = (r_2, \ldots, r_n)$ *satisfying*

$$0 \leqslant r_2 T^{-1} + \lambda_1 < \varepsilon.$$

Then $\mathcal{M}_3(\psi)$ *is null.*

PROOF. Let $\mathfrak{I}_n(N; \mathbf{r}, \mathbf{a})$, where now $\mathbf{a} = (a_{n-1}, \ldots, a_1)$, be the class of polynomials $P \in \mathfrak{I}_n$ of the form $P(t) = Nt^n + a_{n-1}t^{n-1} + \cdots + a_1 t + a_0$ and with root distribution vector $\mathbf{r}$. Now define

$$\widetilde{B}(\alpha_1) = \{t \in (-3,3) \colon |t - \alpha_1| < 2^{-n-1}(n+1)^{-1}|P'(\alpha_1)|^{-1}\}.$$

Again as in (2.33), develop $P(t)$ as a Taylor series on the interval $\widetilde{B}(\alpha_1)$ and estimate each term using the inequality $|P^{(j)}(t)| < j!2^n(n+1)N$ for all $t \in (-3,3)$. Then since $\varepsilon < 1/2$,

$$\max\{|P^{(j)}(\alpha_1)(t-\alpha_1)^j(j!)^{-1}| \colon j = 1, \ldots, n\} < (n+1)^{-1}2^{-n-1},$$

which implies $|P(t)| < 2^{-n-1}$. Now note that $\widetilde{B}(\alpha_1)$ and $\widetilde{B}(\lambda_1)$ are disjoint for any two inequivalent polynomials P, Q from the class $\mathfrak{I}_n(N, \mathbf{r}, \mathbf{a})$ since otherwise the constant difference $P(0) - Q(0)$ satisfies

$$1 \leqslant |R(t)| = |P(t) - Q(t)| = |P(0) - Q(0)| \leqslant 2^{-n}.$$

Thus there are no inessential intervals. It follows that

$$\sum_{P \in \mathfrak{I}_n(N,\mathbf{r},\mathbf{a}))} |\widetilde{B}(\alpha_1)| \leqslant 6.$$

Since the measure $|B(\alpha_1)|$ of the set $B(\alpha_1)$ defined in (2.37) satisfies

$$|B(\alpha_1)| \leqslant |\widetilde{B}(\alpha_1)|\, \psi(N)\, 2^{2n+1}\,(n+1),$$

we have

$$\sum_{P \in \mathfrak{I}_n(N,\mathbf{r})} |B(\alpha_1)| \ll \sum_{\mathbf{a}} \sum_{P \in \mathfrak{I}_n(N,\mathbf{r},\mathbf{a})} |\widetilde{B}(\alpha_1)| \psi(N) \ll \sum_{\mathbf{a}} \psi(N) \ll N^{n-1}\psi(N).$$

Summing over N and $\mathbf{r}$ gives by Cantelli's lemma that the set of t lying in infinitely many $\widetilde{B}(\alpha_1)$ is null and this completes the proof. □

Combining these propositions completes the proof of Theorem 2.1.

2.4.6. The curve $\mathscr{V}$. In order to establish that the normal rational curve $\mathscr{V}$ is a Groshev type manifold for convergence, it remains to show that the convergence of (2.7) implies that $\mathscr{L}(\mathscr{V};\psi)$ is null for each finite interval I. In view of Lemma 1.5, it suffices to show that $\mathfrak{M}(\psi_K)$ is null for $\psi_K(q) = \psi(Kq)$ when $K < 1$. Now if the sum (2.7) converges, then it can be shown by modifying Lemma 5 in [10] that there exists a (slowly decreasing) function $\widetilde{\psi}$ such that $\widetilde{\psi}(q) \geqslant \psi(q)$ for each positive integer q, such that the sum $\sum_q q^{n-1}\widetilde{\psi}(q)$ converges and such that $\widetilde{\psi}(q)/\widetilde{\psi}(Kq) \geqslant c_K$ for some constant $c_K > 0$. Thus $\mathfrak{M}(\widetilde{\psi}) \subseteq \mathfrak{M}(\widetilde{\Psi})$, where $\widetilde{\Psi}(q) = \widetilde{\psi}(q)/c_K \geqslant \widetilde{\psi}(Kq) = \widetilde{\psi}_K(q)$, whence $\mathfrak{M}(\widetilde{\psi}_K) \subseteq \mathfrak{M}(\widetilde{\Psi})$. But $\widetilde{\Psi}$ is decreasing and moreover

$$\sum_q q^{n-1}\widetilde{\Psi}(q) = \frac{1}{c_K}\sum_q q^{n-1}\widetilde{\psi}(q) < \infty,$$

whence $\mathfrak{M}(\widetilde{\Psi})$ and $\mathfrak{M}(\widetilde{\psi}_K)$ are null by the remarks following Theorem 2.1. But $\widetilde{\psi}_K(q) \geqslant \psi_K(q)$ whence $\mathfrak{M}(\psi_K) \subseteq \mathfrak{M}(\widetilde{\psi}_K)$ and so $\mathfrak{M}(\psi_K)$ is null.

2.5. Higher dimensional manifolds

We return to considering the measure of the sets $\mathscr{L}(M;\psi)$ and $\mathscr{S}(M;\psi)$ discussed in §1.5 when M is a smooth m-dimensional manifold. The position for C^3 manifolds when $m \geqslant 2$ is more satisfactory and such manifolds are of Khintchine and Groshev type manifolds for convergence [93] under a mild geometric restriction which requires that locally the manifold is 'bowl-shaped' or 'convex' with respect to two directions. This condition is independent of the choice of coordinates for the manifold, although the proof involves a specific choice which essentially brings the local parametrisation to diagonal form.

2.5.1. A curvature condition. We take M to be a C^3 submanifold of dimension m, embedded in $\mathbb{R}^n$ (so that M has codimension $k = n - m$). The manifold M will be called *2-convex* at the point $\xi \in M$ if for any unit vector $\nu \in T_\xi M^\perp$, at least two of the principal curvatures $\kappa_i(\xi,\nu)$, $i = 1,\dots,m$ (see §1.4.7), are non-zero and have the same sign (this is condition K1 in [93]). When the manifold is a surface in $\mathbb{R}^3$, the condition reduces to the Gaussian curvature $\kappa_1\kappa_2$ being positive and the surface being bowl-shaped in a neighbourhood of ξ. The non-vanishing of a principle curvature for any unit vector in $T_\xi M^\perp$ means that two dimensional convexity excludes the possibility of M lying in a proper Euclidean subspace of $\mathbb{R}^n$ and so is connected with the embedding dimension of M.

Recall from Chapter 1, §1.4.7 that the principal curvatures at $\xi \in M$ with respect to the unit normal $\nu \in T_\xi M^\perp$ have the same sign as the eigenvalues of the matrix $(\nu \cdot \partial^2\theta(u)/\partial u_i\partial u_j)$, where $\theta\colon U \to M$ is a parametrisation with $\theta(u) = \xi$. Recall too that as we are considering measure, we can work locally without loss of generality. Accordingly we consider

$$M_U = \{x \in \mathbb{R}^n : x = \theta(u), u \in U\} = \theta(U),$$

where U is a sufficiently small hypercube and the parametrisation $\theta\colon U \to \mathbb{R}^n$ is C^3. Passing to the Monge form, we consider

$$M_U = \{(u, \varphi(u))\colon u \in U\} = \{(u_1, \dots, u_m, \varphi_1(u), \dots, \varphi_k(u))\colon u \in U\},$$

where the ordinate function $\varphi\colon U \to \mathbb{R}^k$ is also C^3. Recall from (1.21) that we can suppose

$$\sup\{|\partial\varphi_j(u)/\partial u_i|\colon u \in U, i = 1, \dots, m, j = 1, \dots, k\} = K < \infty. \tag{2.41}$$

We study the sets $L(\psi)$ and $S(\psi)$. The set $L(\psi)$ can be written as

$$L(\psi) = \{u \in U\colon u \in B_{\psi(|\mathbf{q}|)}(\mathbf{q}) \text{ for infinitely many } \mathbf{q} \in \mathbb{Z}^n\},$$

where $B_\delta(\mathbf{q}) = \{u \in U\colon \|\mathbf{q}\cdot(u, \varphi(u))\| < \delta\}$, and similarly for $S(\psi)$ (see §1.5.1 above). Its Lebesgue measure is obtained by estimating the measure of elements $B_\delta(\mathbf{q})$ or $B_\delta(q)$ in their natural covers and applying the Cantelli's lemma (measure zero) or a more general form of the Borel lemma (full measure). The sets $L(\psi)$ and $S(\psi)$ depend on the choice of parametrisation θ and domain U but reference to them will not be expressed.

2.5.2. Integral estimates. Estimates for $|B_\delta(\mathbf{q})|$ are obtained from estimates for exponential integrals.

LEMMA 2.16. *Let $U \subset \mathbb{R}^m$ be an open set and let $f\colon U \to \mathbb{R}$ be a C^3 function. If for some α the inequality*

$$\Big|\int_U e(j\,f(u))\,du\Big| \leqslant \alpha\,|j|^{-1} \tag{2.42}$$

holds for all non-zero integers j, then for $\delta \in [0, 1/2]$,

$$\Big|\{u \in U\colon |\langle f(u)\rangle| < \delta\}\Big| = 2\delta\,|U| + O(\alpha\,\delta\,|\log\delta|).$$

PROOF. The function $\chi_{(-\delta,\delta)}(\langle t\rangle)$ is 1-periodic and so has a Fourier series representation given by

$$\chi_{(-\delta,\delta)}(\langle t\rangle) \sim \sum_{j\in\mathbb{Z}} a_j(\delta) e(jt),$$

where $a_j(\delta) = \int_0^1 \chi_{(-\delta,\delta)}(\langle t\rangle)e(ijt)dt$. Hence $a_0(\delta) = 2\delta$ and for $j \neq 0$,

$$|a_j(\delta)| = \left|\frac{\sin 2\pi j\delta}{\pi j}\right| \leqslant \min\left\{2\delta, \frac{1}{\pi|j|}\right\},$$

so that the integral

$$\int_U \chi_{(-\delta,\delta)}(\langle f(u)\rangle)\,du = 2\delta\,|U| + \sum_{j\neq 0} a_j(\delta)\int_U e(jf(u))\,du.$$

But by (2.42),

$$\left|\sum_{j\neq 0} a_j(\delta)\int_U e(j\,f(u))\,du\right| \leqslant \sum_{1\leqslant |j|\leqslant 1/2\pi\delta} (2\delta)\,\frac{\alpha}{|j|} + \frac{1}{\pi}\sum_{|j|>1/2\pi\delta} \frac{\alpha}{|j|^2}$$
$$\ll \alpha\delta|\log\delta| + \alpha\int_{\pi/2\delta}^{\infty}\frac{dx}{x^2}$$
$$\ll \alpha\delta(|\log\delta|+1) \ll \alpha\delta|\log\delta|.$$

Hence

$$\big||\{u\in U\colon \|f(u)\|<\delta\}| - 2\delta|U|\big| = \Big|\sum_{j\neq 0} a_j(\delta)\int_U e(jf(u))\,du\Big| \ll \alpha\delta|\log\delta|,$$

as claimed. □

The next two results relate derivative properties to estimates for exponential integrals and are Lemmas 2.2 and 2.3 respectively in [93].

LEMMA 2.17. *Let $h\colon I\to\mathbb{R}$ be a C^2 map of the interval I such that for some positive constants β, c, the inequalities*

$$|h'(u)|\geqslant\beta,\quad |h''(u)|\leqslant c\beta$$

hold. Then for any subinterval J of I,

$$\Big|\int_J e(h(u))\,du\Big| \leqslant \frac{(1+c|J|)}{\pi\beta}.$$

This is essentially Lemma 4.2 of [217] and follows from integrating by parts.

LEMMA 2.18. *Let $h\colon\mathbb{R}^2\to\mathbb{R}$ be a C^3 map. Suppose that the partial derivatives $\partial^2 h(u_1,u_2)/\partial u_1^2$, $\partial^2 h(u_1,u_2)/\partial u_2^2$ have the same sign. Suppose further that for all $(u_1,u_2)\in\mathbb{R}^2$, the inequalities*

$$\left|\frac{\partial^2 h(u_1,u_2)}{\partial u_1^2}\right|,\ \Big|\frac{\partial^2 h(u_1,u_2)}{\partial u_2^2}\Big| \geqslant\beta,\quad \Big|\frac{\partial^2 h(u_1,u_2)}{\partial u_1\partial u_2}\Big|\leqslant\beta/2$$

and

$$\Big|\frac{\partial^3 h(u_1,u_2)}{\partial u^i\partial u_2^j}\Big|\leqslant c\beta,\ i+j=3,$$

hold for some constants c,β. Then for any planar convex set A,

$$\Big|\int_A e(h(u_1,u_2))\,du_1\,du_2\Big| \leqslant \frac{(1+c|A|)}{\pi\beta}.$$

The proof relies on the function $f(u_1) = \partial h(u_1, u_2)/\partial u_1$, regarded as a function of a single real variable u_1 with u_2 fixed, having a derivative bounded away from 0, so that for each u_2 there is a unique $\tilde{u}_1 = \tilde{u}_1(u_2)$ such that $f(\tilde{u}_1) = 0$. The other conditions guarantee that the Hessian $(\partial^2 h(u_1, u_2)/\partial u_i \partial u_j)$ is invertible, so that by the implicit function theorem, $\tilde{u}_1(u_2)$ is a C^2 function of u_2, whence the function $\tilde{h}\colon \mathbb{R} \to \mathbb{R}$ given by $\tilde{h}(u_2) = \tilde{h}(\tilde{u}_1(u_2), u_2)$ is C^2 and $\tilde{h}''(u_2) \geqslant c\beta$.

2.5.3. Two auxiliary functions. We now introduce two auxiliary functions Φ_ω and g. Let ω be a non-zero vector in $\mathbb{R}^k$. Define the function $\Phi_\omega\colon U \to \mathbb{R}$ by

$$\Phi_\omega(u) = \omega \cdot \varphi(u).$$

The Hessian of Φ_ω given by

$$\operatorname{Hess} \Phi_\omega(u) = \left(\omega \cdot \frac{\partial^2 \varphi(u)}{\partial u_i \partial u_j}\right) = \left(\sum_{r=1}^{k} \omega_r \frac{\partial^2 \varphi_r(u)}{\partial u_i \partial u_j}\right).$$

is a continuous function of ω as φ is C^3.

Normalise the vector $Y = \sum_{r=1}^{k} \omega_r Y_r \in T_\xi M^\perp$, where $Y_1, \ldots, Y_k$ are a basis for $T_x M^\perp$ (see (1.26)), to $\nu = Y/|Y|_2$, where

$$\begin{aligned} |Y|_2^2 &= \sum_{r,s=1}^{k} \omega_r \omega_s (\operatorname{grad} \varphi_r(u), -e_r) \cdot (\operatorname{grad} \varphi_s(u), -e_s) \\ &\leqslant \sum_{r,s=1}^{k} |\omega_r||\omega_s| \left(\sum_{i=1}^{m} \left|\frac{\partial \varphi_r(u)}{\partial u_i} \frac{\partial \varphi_s(u)}{\partial u_i}\right| + \delta_{rs}\right). \end{aligned}$$

By the choice of U,

$$|\omega|^2 \leqslant |\omega|_2^2 \leqslant |Y|_2^2 \leqslant mK^2|\omega|_1^2 + |\omega|_2^2 \leqslant k(mK^2+1)|\omega|^2. \tag{2.43}$$

But by (1.26)

$$\begin{aligned} \left(\nu \cdot \frac{\partial^2 \theta(u)}{\partial u_i \partial u_j}\right) &= \left(\frac{Y}{|Y|_2} \cdot \frac{\partial^2 \theta(u)}{\partial u_i \partial u_j}\right) = \left(\sum_{r=1}^{k} \frac{-\omega_r}{|Y|_2} e_r \cdot \frac{\partial^2 \varphi(u)}{\partial u_i \partial u_j}\right) \\ &= -\frac{1}{|Y|_2} \operatorname{Hess} \Phi_\omega(u), \end{aligned}$$

so that the eigenvalues $\lambda_1, \ldots, \lambda_m$ say of $\operatorname{Hess} \Phi_\omega(u)$ are given by

$$\lambda_j = -|Y|_2 \gamma_j \kappa_j, \quad j = 1, \ldots, m,$$

where $\gamma_1, \ldots, \gamma_m$ are the (positive) eigenvalues of the metric matrix G (1.23) given by

$$G = (g_{ij}) = \left(\frac{\partial \varphi(u)}{\partial u_i} \cdot \frac{\partial \varphi(u)}{\partial u_j}\right)$$

and $\kappa_1, \ldots, \kappa_m$ are the principal curvatures at $\xi = (u, \varphi(u))$ with respect to $\nu = Y/|Y|_2$. Thus M is 2-convex at $\xi = (u, \varphi(u))$ if and only if for each non-zero $\omega \in \mathbb{R}^k$, $\operatorname{Hess} \Phi_\omega(u)$ has at least two non-zero eigenvalues of the same sign.

In order to discuss Hess Φ_ω in a neighbourhood of u, we introduce a new variable u^* with $|u-u^*|$ small and function Φ^*_ω given by $\Phi^*_\omega(u^*) = \Phi_\omega(u-u^*)$. For simplicity, we drop the * and consider $\Phi_\omega(u)$ for $|u|$ small. Thus M_U is contained in a hypercube centred at 0 of sidelength comparable with ε.

Let $\tilde{\varepsilon} > 0$ be sufficiently small and fixed and let $\varepsilon \in (0, \tilde{\varepsilon})$. Then

$$U = (-\varepsilon, \varepsilon)^m \subset \widetilde{U} = (-\tilde{\varepsilon}, \tilde{\varepsilon})^m.$$

Since φ is C^3, by Taylor's theorem, for fixed $\omega \in \mathbb{R}^k$ and for each $u \in \widetilde{U}$,

$$\Phi_\omega(u) = \Phi_\omega(0) + \operatorname{grad}\Phi_\omega(0)\cdot u + \frac{1}{2}Q_\omega(u) + R(u),$$

where $Q_\omega(u) = u \operatorname{Hess}\Phi_\omega(0)\, u^T = \sum_{i,j}(\operatorname{Hess}\Phi_\omega(0))_{ij}u_i u_j$ and $R(u)$ is the remainder term. The function $\Phi_\omega\colon U \to \mathbb{R}$ is extended to a C^3 function $\widetilde{\Phi}_\omega\colon \mathbb{R}^m \to \mathbb{R}$ defined on $\mathbb{R}^m$ and given by

$$\widetilde{\Phi}_\omega(u) = \begin{cases} \Phi_\omega(0) + \operatorname{grad}\Phi_\omega(0)\cdot u + \frac{1}{2}Q_\omega(u) + b(|u|/\varepsilon)R(u), & u \in \widetilde{U}, \\ \Phi_\omega(0) + \operatorname{grad}\Phi_\omega(0)\cdot u + \frac{1}{2}Q_\omega(u), & u \notin \widetilde{U}, \end{cases}$$

by means of a smooth 'bump' function $b\colon \mathbb{R} \to \mathbb{R}$, where $b(t) = 1$, $0 \leqslant |t| \leqslant 1$, $b(t) = 0$, $|t| \geqslant \tilde{\varepsilon}/\varepsilon$ and $b(t)$ is smooth for $1 \leqslant |t| \leqslant \tilde{\varepsilon}/\varepsilon$. Clearly $\widetilde{\Phi}_\omega = \Phi_\omega$ on U. It follows from continuity and compactness that given $\eta > 0$, there exists an $\varepsilon > 0$ such that

$$|\operatorname{Hess}\widetilde{\Phi}_\omega(u) - \operatorname{Hess}\widetilde{\Phi}_\omega(0)| < \eta$$

for $(u, \omega) \in U \times \mathbb{R}^k$. Let T be an orthogonal matrix which diagonalises the matrix Hess $\widetilde{\Phi}_\omega(0)$. Then it is readily verified that

$$|T\Big(\operatorname{Hess}\widetilde{\Phi}_\omega(u) - \operatorname{Hess}\widetilde{\Phi}_\omega(0)\Big)T^{-1}| \leqslant m^2\eta.$$

Recall from (1.35) that $\mathbf{q} = (\mathbf{q}^{(1)}, \mathbf{q}^{(2)}) \in \mathbb{Z}^m \times \mathbb{Z}^k$. Define the function $g\colon U \to \mathbb{R}$ by

$$g(u) = \mathbf{q}\cdot\theta(u) = \mathbf{q}^{(1)}\cdot u + \mathbf{q}^{(2)}\cdot\varphi(u) = \mathbf{q}^{(1)}\cdot u + \Phi_{\mathbf{q}^{(2)}}(u). \tag{2.44}$$

The domain of the function $g\colon U \to \mathbb{R}$ can be extended to $\mathbb{R}^m$ by defining the function $\tilde{g}\colon \mathbb{R}^m \to \mathbb{R}$ as

$$\tilde{g}(u) = \mathbf{q}^{(1)}\cdot u + \widetilde{\Phi}_{\mathbf{q}^{(2)}}(u).$$

This is a standard device which prevents problems at the boundary of U. For simplicity we will not distinguish between g and its extension $\tilde{g}$. Note that given $\eta' > 0$, there exists an $\varepsilon > 0$ such that for each $u \in U = (-\varepsilon, \varepsilon)^m$,

$$|\operatorname{Hess} g(u) - \operatorname{Hess} g(0)| = |\operatorname{Hess}\widetilde{\Phi}_{\mathbf{q}^{(2)}}(u) - \operatorname{Hess}\widetilde{\Phi}_{\mathbf{q}^{(2)}}(0)| < \frac{\eta'}{2m^2}. \tag{2.45}$$

LEMMA 2.19. *Let U be a sufficiently small hypercube and $\theta\colon U \to \mathbb{R}^n$ be C^3 and one-to-one. Suppose that for each $u \in U$ and each unit vector normal to $T_{\theta(u)}\theta(U)$, at least two principal curvatures of the manifold $\theta(U)$ are of the same sign and at least 2δ in modulus for some $\delta > 0$. Then for any non-zero vector $\mathbf{q} \in \mathbb{Z}^n$,*

$$\left|\int_U e(\mathbf{q}\cdot\theta(u))\,du\right| \ll 1/|\mathbf{q}|.$$

The hypotheses imply that $M = \theta(U)$ is 2-convex.

PROOF. By hypothesis, $U = I(\varepsilon)^m$ where $I(\varepsilon)$ is an interval of sufficiently small length ε. We pass to the Monge parametrisation and express each point in M in the form $\theta(u) = (u, \varphi(u))$, where $\varphi\colon U \to \mathbb{R}^k$ is C^3. From (2.44), for each $u \in U$

$$\frac{\partial g(u)}{\partial u_i} = q_i + \mathbf{q}^{(2)}\cdot\frac{\partial\varphi(u)}{\partial u_i}, \quad i = 1, \ldots, m.$$

First suppose that $|\mathbf{q}^{(2)}| \leqslant |\mathbf{q}|/(2kK)$, where $|\partial\phi_j(u)/\partial u_i| \leqslant K$ (see (2.41)), so that the sets $\{x \in M_U\colon \langle\mathbf{q}\cdot x\rangle = 0\}$ are transverse to M. By relabelling we can take q_1 to have maximum modulus, *i.e.*, $|q_1| = |\mathbf{q}|$. Then for each $u \in U$,

$$\left|\frac{\partial g(u)}{\partial u_1}\right| \geqslant |q_1| - \left|\mathbf{q}^{(2)}\cdot\frac{\partial\varphi(u)}{\partial u_1}\right| \geqslant \frac{1}{2}|\mathbf{q}|. \tag{2.46}$$

Moreover by continuity, $|\partial^2 g(u)/\partial u_1^2| = |\mathbf{q}^{(2)}\cdot\partial^2\varphi(u)/\partial u_1^2| \ll |\mathbf{q}|$. Now since $U = I(\varepsilon)^m$, by Fubini's theorem,

$$\int_U e(\mathbf{q}\cdot\theta(u))\,du = \int_{I(\varepsilon)^{m-1}}\left(\int_{I(\varepsilon)} e(g(u_1,u'))\,du_1\right)du',$$

where $u = (u_1, u') \in I(\varepsilon)\times I(\varepsilon)^{m-1}$. Hence by Lemma 2.17 with $h(u_1) = g(u_1, u')$ and $\beta = |\mathbf{q}|/2$ applied to the integral over $I(\varepsilon)$, it follows that

$$\left|\int_U e(\mathbf{q}\cdot\theta(u))\,du\right| \ll |\mathbf{q}|^{-1}$$

and the lemma holds in this case.

Next suppose that $|\mathbf{q}^{(2)}| > |\mathbf{q}|/(2kK)$, corresponding to the more difficult case when the sets $\{x \in M_U\colon \langle\mathbf{q}\cdot x\rangle = 0\}$ are near to being tangential to M_U. Fix $u \in U$ and again for convenience change the coordinates so that $\xi = \theta(0) = 0$. Then $U = (-\varepsilon, \varepsilon)^m$. Recall that

$$\operatorname{Hess} g(0) = \left(\mathbf{q}^{(2)}\cdot\frac{\partial^2\varphi(0)}{\partial u_i\partial u_j}\right) = -|Y|_2\left(\nu\cdot\frac{\partial^2\theta(0)}{\partial u_i\partial u_j}\right), \tag{2.47}$$

where

$$Y = \sum_{r=1}^{k} q_r^{(2)} Y_r = \sum_{r=1}^{k} q_{m+r} Y_r \in T_x M^{\perp} \quad \text{and} \quad \nu = \gamma/|\gamma|_2 \in T_x M^{\perp}.$$

Make the orthogonal transformation T of the coordinates in $\mathbb{R}^m$ where T diagonalises $\operatorname{Hess} g(0)$. For simplicity a new notation for the transformed coordinates Tu and functions will not be introduced. Thus we take

$$\operatorname{Hess} g(0) = \operatorname{diag}(\lambda_1, \lambda_2, \ldots, \lambda_m).$$

By hypothesis at least two principal curvatures are of the same sign and at least 2δ in modulus. By (2.47) and by relabelling the coordinates if necessary we can take $|\lambda_1|, |\lambda_2| \geqslant 2|Y|_2\,\gamma\,\delta$ where $\gamma = \max\{\gamma_1, \gamma_2\}$ and γ_i are the (positive) eigenvalues of the metric matrix G of the manifold at ξ. Now from (2.45), for each $u \in U$, each element $\partial^2 g(u)/\partial u_i \partial u_j$ of $\operatorname{Hess} g(u)$ is close to the corresponding element of the diagonal matrix $\operatorname{Hess} g(0)$ for $i, j = 1, 2, \ldots, m$. In particular the diagonal element $\partial^2 g(u)/\partial u_i^2$ is close to λ_i for each $i = 1, \ldots, m$. Choose

$$\eta' = \frac{|Y|_2 \gamma \delta}{(4k(mK^2+1))^{1/2}} \leqslant \frac{|Y|_2 \gamma \delta}{2}.$$

Then for ε sufficiently small, (2.45) and (2.43) imply that

$$\left|\frac{\partial^2 g(u)}{\partial u_1^2}\right|, \left|\frac{\partial^2 g(u)}{\partial u_2^2}\right| > |Y|_2\,\gamma\delta \geqslant |\mathbf{q}^{(2)}|\,\gamma\delta$$

and

$$\left|\frac{\partial^2 g(u)}{\partial u_1 \partial u_2}\right| < \frac{|Y|_2\,\delta\gamma}{(4k(mK^2+1))^{1/2}} \leqslant \frac{|\mathbf{q}^{(2)}|\,\delta}{2}.$$

By the definition of g,

$$\left|\frac{\partial^3 g(u)}{\partial u_1 \partial u_2 \partial u_3}\right| \ll |\mathbf{q}^{(2)}| \ll |\mathbf{q}|.$$

Again by Fubini's theorem,

$$\int_U e(\mathbf{q}\cdot\theta(u))\,du = \int_{I(\varepsilon)^{m-2}} \left(\int_{I(\varepsilon)^2} e(g(u_1, u_2, u''))\,du_1\,du_2\right) du'',$$

where now $u = (u_1, u_2, u'') \in I(\varepsilon)^2 \times I(\varepsilon)^{m-2}$. The square $I(\varepsilon)^2$ is convex, so by Lemma 2.18 with $A = I(\varepsilon)^2$ and $h(u_1, u_2) = g(u_1, u_2, u'')$ applied to the above estimates, the inequalities

$$\left|\int_U e(\mathbf{q}\cdot\theta(u))\,du\right| \ll \varepsilon^{m-2}\left|\int_{I(\varepsilon)^2} e(g(u_1, u_2, u''))\,du_1\,du_2\right| \ll |\mathbf{q}|^{-1},$$

hold. This completes the lemma. □

We can now prove the following result.

THEOREM 2.20. *Let M be a C^3, m-dimensional submanifold embedded in $\mathbb{R}^n$. Suppose $m \geqslant 2$ and M is 2-convex almost everywhere. Then M is a Khintchine-Groshev manifold for convergence.*

PROOF. From the hypothesis, the set of points ξ in M for which no two of the principal curvatures $\kappa(\xi, \nu)$ are of the same sign and non-zero for every unit normal ν in $T_\xi M^\perp$ is null. Hence this set can be covered by a countable number of open sets in M with arbitrarily small total (induced) measure. The complement in M is a countable union of closed sets in which at least two principal curvatures are of the same sign and bounded away from 0. Thus we need only consider parametrisations $\theta\colon U \to \mathbb{R}^n$ where U is a suitably small hypercube such that for each $u \in U$, the modulus of at least two principal curvatures of $\theta(u)$ is at least 2δ.

Let j be a non-zero integer and $\mathbf{q}$ a non-zero vector in $\mathbb{Z}^n$. Then it follows from Lemma 2.19 that

$$\left|\int_U e(j\,\mathbf{q}\cdot\theta(u))\,du\right| \ll (|j||\mathbf{q}|)^{-1}.$$

Hence it follows from Lemma 2.16 that the set $B_\delta(\mathbf{q}) = \{u \in U\colon |\langle \mathbf{q}\cdot\theta(u)\rangle| < \delta\}$ has measure

$$|B_\delta(\mathbf{q})| = 2\delta|U| + O\big(\delta|\log\delta|/|\mathbf{q}|\big). \tag{2.48}$$

The log term in the estimate prevents a direct application of Cantelli's lemma and we have to make a little detour.

Let $\psi(r), r = 1, 2, \ldots$, be a sequence of positive numbers such that the sum $\sum_r \psi(r)^n$ is convergent. Introduce the auxiliary sequence $\psi^*(r)$, $r = 1, 2, \ldots$, given by $\psi^*(1) = 1/2$ and

$$\psi^*(r) = \max\{\psi(r),\, r^{-n-1}\}, \quad r = 2, 3, \ldots.$$

Then $\psi^*(r) \geqslant \psi(r)$, so that $L(\psi) \subseteq L(\psi^*)$ and

$$|\log\psi^*(r)| = \min\{|\log\psi(r)|,\, (n+1)\log r\} \ll \log r$$

when $r \geqslant 2$. It follows that

$$\sum_{\mathbf{q}\neq 0}\psi(|\mathbf{q}|) \leqslant \sum_{\mathbf{q}\neq 0}\psi^*(|\mathbf{q}|) \ll \sum_{\mathbf{q}\neq 0}\psi(|\mathbf{q}|) + \sum_{\mathbf{q}\neq 0}\frac{1}{|\mathbf{q}|^{n+1}}$$

and so the sum $\sum_q \psi^*(q)q^{n-1}$ is convergent if and only if $\sum_q \psi(q)q^{n-1}$ (see (2.7)) is convergent. By definition, $\mathscr{C} = \{B_{\psi^*(|\mathbf{q}|)}(\mathbf{q})\colon \mathbf{q} \in \mathbb{Z}^n \setminus \{0\}\}$ is a natural cover for $L(\psi^*)$ and hence by (2.48)

$$\sum_{\mathbf{q}} |B_{\psi^*(|\mathbf{q}|)}(\mathbf{q})| \ll \sum_{\mathbf{q}}\psi^*(|\mathbf{q}|) + \sum_{\mathbf{q}}\frac{\psi^*(|\mathbf{q}|)\,|\log\psi^*(|\mathbf{q}|)|}{|\mathbf{q}|} < \infty.$$

Thus by Cantelli's lemma, $|L(\psi^*)| = 0$, whence $|L(\psi)| = 0$. It follows that $|\mathscr{L}(M;\psi)|_M = 0$ and that M is a Groshev type manifold for convergence.

By adapting a result of E. Kovalevskaya ([210, p. 107, Lemma 8], see also [89, §5]), these ideas can be used to show that M is a Khintchine type manifold for convergence as well. Let $\psi(q)$, $q = 1, 2, \ldots$, be a sequence of positive numbers such that (2.6) is convergent. For each positive integer q, let

$$0B_\delta(q) = \{u \in U\colon |\langle q\theta(u)\rangle| < \delta\}.$$

Then the family $\{B_{\psi(q)}(q)\colon q\in\mathbb{N}\}$ is a natural cover for $S(\psi)$ and by [210, p. 107],

$$|B_{\psi(q)}(q)| \ll \psi(q)^n \sum_{\substack{|\mathbf{r}|\leqslant 1/\psi(q)\\ \mathbf{r}\in\mathbb{Z}^n}} \left|\int_U e(\mathbf{r}\cdot q\theta(u))\,du\right| \ll \psi(q)^n\left(1+\sum_{\substack{1\leqslant|\mathbf{r}|\leqslant 1/\psi(q)\\ \mathbf{r}\in\mathbb{Z}^n}}\frac{1}{q|\mathbf{r}|}\right)$$

by Lemma 2.19. Hence $|B_{\psi(q)}(q)| \ll \psi(q)^n\left(1+q^{-1}\psi(q)^{-n+1}\right)$. Next define the auxiliary sequence $\psi^*(q)$, $q=1,2,\ldots$, by $\psi^*(1)=1/2$ and

$$\psi^*(q)=\max\{\psi(q),\, q^{-\eta-1/n}\},\quad q=2,3,\ldots,$$

for some η satisfying $0<(n-1)\eta<1/n$. Then as before, $\psi(q)\leqslant\psi^*(q)$ when $q\geqslant 2$, and

$$q^{-1}\psi^*(q)^{-n+1}\leqslant q^{\eta(n-1)-1/n}.$$

Thus $S(\psi)\subseteq S(\psi^*)$ and the series $\sum_{q=1}^\infty\psi^*(q)^n$ is convergent if and only if the series (2.6) is convergent. The choice of η implies that $|B_{\psi^*(q)}(q)|\ll\psi^*(q)^n$. It follows that

$$\sum_{q=1}^\infty |B(\psi^*(q))| \ll \sum_{q=1}^\infty \psi^*(q)^n<\infty$$

and so by Cantelli's lemma, $|S(\psi^*)|=0$, whence $|S(\psi)|=0$ and so $|\mathscr{S}(M;\psi)|_M=0$, as required. $\square$

Since the sum $\sum_r r^{-n-\varepsilon}$ converges for any $\varepsilon>0$, putting $\psi(r) = r^{-n-\varepsilon}$ in the definition of Groshev type for convergence in §2.3 gives immediately that a C^3 manifold which is 2-convex almost everywhere is extremal. Of course Kleinbock and Margulis have proved more general manifolds extremal (see §2.2 above). The case when the sum diverges is more difficult but it is shown in [89] that a smooth 2-convex manifold is also of Groshev type for divergence, *i.e.,* when the series $\sum_r\psi(r)\,r^{n-1}$ diverges, $\mathscr{L}(M\,;\psi)$ has full measure. This requires estimating the measure of the intersection $B_\delta(\mathbf{q})\cap B_{\delta'}(\mathbf{q}')$ in terms of $|B_\delta(\mathbf{q})|\,|B_{\delta'}(\mathbf{q}')|$.

2.6. Notes

§2.2 Mahler's conjecture is closely related to his classification of the real and complex numbers in terms of $\omega_n(t)$ (details are in [115], [208]). Sprindžuk proved the conjecture for both the real case and F. Kasch's refinement [130] in the complex case; in addition he obtained analogues for p-adic fields and the field of formal power series. He also proved a partial Groshev type convergence result for the complex curve $\{(z^m,z^n)\colon z\in\mathbb{C}\}$, where $1\leqslant m<n$ [208]. I. M. Morozova [165] has obtained a sharper result for certain leading lacunary integer polynomials of a complex variable.

The definition of extremality could equally be given in terms of the related sets $\mathcal{L}_v(M)$ and $\mathcal{S}_v(M)$.

Extremal manifolds arise in the study of dynamical systems, for example in connection with averaging over fast variables (see Chapter 7 and [8, p. 161], [96]).

Points $x\in M$ which satisfy (2.5) for infinitely many $\mathbf{q}\in\mathbb{Z}^n$ are called *multiplicatively very well approximable* points since by a modification of Khintchine's transference principle [210, p. 69], the inequality (2.5) is equivalent to the multiplicative form $\prod_{j=1}^n\|q\,x_j\|<q^{-vn}$, $v>1/n$.

Metric results for multiplicative Diophantine approximationare discussed in [108], [205].

Exponential sums and integrals are useful for studying simultaneous approximation on manifolds $M \subset \mathbb{R}^n$ when $\dim M \geqslant n/2$, see [145]. Cassels [58] and Kubilius [146] and others also proved stronger results for the parabola [31], [156], [235]. These methods were also used in [45], [143], [160], [210].

The product $M \times M'$ of two extremal manifolds M, M' is an extremal manifold in $\mathbb{R}^{n+n'}$ [30]. Simultaneous approximation which involves $\max\{\|\mathbf{q} \cdot x\|, \|\mathbf{q} \cdot y\|\}$ is connected with Sprindžuk's conjecture [207], [208] and is considered in [32], [97], [236].

Margulis [154] used the density of certain orbits in a homogeneous space to prove Oppenheim's conjecture that given $\varepsilon > 0$, any indefinite quadratic form $Q(x)$ not a scalar multiple of a rational form satisfies $|Q(\mathbf{q})| < \varepsilon$ for some integer vector $\mathbf{q}$.

A curve satisfying the hypotheses of Schmidt's theorem has no straight segments. A proof of Schmidt's theorem using essential domains is in [210, Chapter 2]. Hausdorff dimension is discussed in Chapters 4 and 5 which contain further references.

Extremality was weakened to ν-extremality in which the exponent $-n - \varepsilon$ was replaced by $-\nu n - \varepsilon$ with $\nu \geqslant 1$ [226]. The extra freedom allowed a variety of results, see for example [17], [36], [186], [187], [209]; some of these have been superseded by Kleinbock and Margulis' proof of Sprindžuk's conjecture [139].

N. I. Markovich [155] has obtained a $((3k+1)/2)$-extremality type of result for the quadratic $a_0 + a_1 t + a_2 t^2$ when the a_j are integers in a real algebraic extension of $\mathbb{Q}$ of degree k.

§**2.3** Using analytic and geometrical techniques based on those of Bernik and Sprindžuk, Beresnevich has proved that smooth manifolds nondegenerate almost everywhere are of Groshev type for convergence [24]. Bernik, Kleinbock and Margulis have proved the same result for both convergence and divergence when the manifolds are analytic; in fact for convergence they prove the stronger multiplicative version [41]. They use flows and Beresnevich's improved regular systems [22, 23].

The conjecture could be extended to systems of linear forms by considering a product $M = \prod_{j=1}^{n} M_j$ of manifolds, where each $M_j \subset \mathbb{R}^m$ is nondegenerate almost everywhere.

Kunrui Yu considered a related multiplicative Khintchine type problem for the parabola in [235] and showed that if $\sum_q \psi^2(q)$ then the set of t for which there exist infinitely many integer solutions of the inequality

$$\|qt\| \, \|qt^2\| < q^3 \psi^4(q)$$

is null. The approximation function was improved to $\psi(q)^2/\log(q)$ by V. I. Mashanov [156].

Bernik [32] has proved Sprindžuk's 'multiplicative' conjecture [207] in which $P(t)$ is replaced by the product $P(t_1) \dots P(t_k)$; F. F. Zheludevich [236] extended the result to the simultaneous approximations in metrics of $\mathbb{R}, \mathbb{C}, \mathbb{Q}_p$. I. R. Dombrovsky [97] obtained a result for the product $P_1(t)P_2(t)$ of two different integral polynomials P_1, P_2 with the first k coefficients in common.

If a manifold is of Khintchine or Groshev type for divergence, it is of interest to know how rapidly the number of solutions grows. Schmidt's [196] very general quantitative refinement of Khintchine's theorem has been used to obtain results for independent variables [28, 196] and for certain manifolds [29, 95, 210]. In the case of convergence, the Hausdorff dimension of the null sets is discussed in Chapters 4 and 5.

§**2.4** By Khintchine's result [136], the set of points t for which the inequality (2.9) holds infinitely often has full measure.

§**2.5.1** The non-vanishing of a principal curvature $\kappa_i(\xi, \nu)$ for each $\nu \in T_\xi M^\perp$ implies that the manifold is geometrically full.

CHAPTER 3

Hausdorff measure and dimension

3.1. Introduction

Hausdorff measure and dimension stem from F. Hausdorff's simple but far-reaching variation [116] of C. Carathéodory's approach to Lebesgue measure [56] (more details are given in the Notes at the end of the chapter). For familiar sets such as the interval, circle, sphere and the plane, the Hausdorff dimension (defined below in §3.3) coincides with the usual notion of dimension and is respectively 1, 1 ,2 and 2. However, an important difference is that *any* set in Euclidean space has a Hausdorff dimension. In particular, null sets have a Hausdorff dimension and this gives a way of discriminating between them. The study of this finer aspect of the metric structure of exceptional sets, which started with Hausdorff's determination of the dimension of the Cantor 'middle third' set, was developed by A. S. Besicovitch and V. Jarník and continues unabated.

Hausdorff measure has been studied intensively and in considerable generality, indeed the theory can be extended to a metric space setting. This tract will be concerned mainly with Borel subsets of submanifolds of Euclidean space and accordingly the treatment of Hausdorff measure and dimension will be in $\mathbb{R}^n$. Fuller treatments and further references can be found in the books of K. Falconer [100, 101], H. Federer [103], P. Mattila [157] and C. A. Rogers [189]. Applications to exceptional sets in number theory are discussed in [115, Chapter 10].

3.2. Hausdorff measure

Hausdorff measure is based on covers. Let E be a set in $\mathbb{R}^n$ and let s be a non-negative real number. Hausdorff's idea was to introduce for each cover $\mathscr{C}$ of E in $\mathbb{R}^n$ the (possibly infinite) sum

$$\ell^s(\mathscr{C}) = \sum_{C\in\mathscr{C}} (\operatorname{diam} C)^s,$$

where $\operatorname{diam} C = \sup\{|x-y|_2 \colon x, y \in C\}$ is the diameter of C. This quantity can be regarded as a generalised length and will be called the *s-length* of the cover $\mathscr{C}$. A *δ-cover* of E is a cover of E by sets, each of diameter at most δ. A δ-cover will usually be denoted by a suffix δ, so that $\mathscr{C}_\delta(E)$ say is a collection of sets, each of diameter at most δ, which cover E. Covers $\mathscr{C}(E)$ of a given set E in $\mathbb{R}^n$ by a restricted collection of sets $C \in \mathscr{C}$, such as balls, cubes or more generally convex sets, turn out to involve no real loss of generality and can be easier to work with.

Let s be a non-negative real number. The quantity

$$\mathcal{H}_\delta^s(E) = \inf \sum_{C\in\mathscr{C}} (\operatorname{diam} C)^s = \inf \ell^s(\mathscr{C}), \tag{3.1}$$

where the infimum is taken over all finite or countable δ-covers $\mathscr{C}$ of E, is a non-negative real number or infinite. It can be verified that $\mathcal{H}_\delta^s$ is an outer measure on $\mathbb{R}^n$ but the proof is omitted since it is essentially the same as that for the better behaved Hausdorff outer measure $\mathcal{H}^s$ about to be introduced. If $0 < \rho < \delta$, then a ρ-cover is a δ-cover as well. Hence

$$\mathcal{H}_\delta^s(E) \leqslant \mathcal{H}_\rho^s(E) \tag{3.2}$$

since on the right hand side the infimum is taken over a smaller choice of covers than on the left hand side and so cannot be smaller. Therefore the limit (which might be infinite)

$$\mathcal{H}^s(E) = \lim_{\delta\to 0} \mathcal{H}_\delta^s(E) \tag{3.3}$$

and $\mathcal{H}^s(E) \geqslant \mathcal{H}_\delta^s(E)$ for each $\delta > 0$. Given a set E and a δ-cover $\mathscr{C}$, it is clear that when $\delta < 1$, $\sum_{C\in\mathscr{C}}(\operatorname{diam} C)^s$ cannot increase with s. Hence by (3.1), $\mathcal{H}_\delta^s(E)$ also cannot increase with s and so by (3.3), $\mathcal{H}^s(E)$ also cannot increase with s. Moreover it is clear from (3.1) that $\mathcal{H}_\delta^s(\emptyset)$ vanishes for each $\delta > 0$, whence $\mathcal{H}^s(\emptyset) = 0$.

If $E \subseteq F$, then any cover of F is a cover for E, so that for each positive δ, $\mathcal{H}_\delta^s(E) \leqslant \mathcal{H}_\delta^s(F)$, whence

$$\mathcal{H}^s(E) \leqslant \mathcal{H}^s(F) \tag{3.4}$$

and the set function $\mathcal{H}^s$ is increasing. This also follows from $\mathcal{H}^s$ being subadditive; the proof of this result is included as an illustration of a useful technique.

LEMMA 3.1. *For each $s \geqslant 0$, $\mathcal{H}^s$ is subadditive.*

PROOF. Let $E_1, E_2, \ldots$ be any finite or countable collection of sets and fix an arbitrary $\varepsilon > 0$. There is no loss in generality in assuming $\mathcal{H}_\delta^s(E_j)$ finite. By definition, for each j there is a δ-cover $\mathscr{C}^{(j)}$ of E_j such that

$$\ell^s(\mathscr{C}^{(j)}) = \sum_{C\in\mathscr{C}^{(j)}} (\operatorname{diam} C)^s < \mathcal{H}_\delta^s(E_j) + \frac{\varepsilon}{2^j} \leqslant \mathcal{H}^s(E_j) + \frac{\varepsilon}{2^j}$$

by (3.3) and by (3.2). Evidently the sets $C \in \mathscr{C}^{(j)}$, $j = 1, 2, \ldots$, form a δ-cover for the set $\bigcup_{j=1}^\infty E_j$. Adding the above inequalities for $j = 1, 2, \ldots$, we obtain

$$\sum_{j=1}^\infty \ell^s(\mathscr{C}^{(j)}) = \sum_{j=1}^\infty \sum_{C\in\mathscr{C}^{(j)}} (\operatorname{diam} C)^s \leqslant \sum_{j=1}^\infty \mathcal{H}^s(E_j) + \varepsilon.$$

Because $\mathcal{H}_\delta^s$ is defined as an infimum (3.1),

$$\mathcal{H}_\delta^s(\bigcup_j E_j) \leqslant \sum_{j=1}^\infty \sum_{C\in\mathscr{C}^{(j)}} (\operatorname{diam} C)^s$$

and by the definition of $\mathcal{H}^s$, for δ sufficiently small, $\mathcal{H}^s_\delta(\bigcup_j E_j) > \mathcal{H}^s(\bigcup_j E_j) - \varepsilon$, so that

$$\mathcal{H}^s(\bigcup_j E_j) \leqslant \mathcal{H}^s_\delta(\bigcup_j E_j) + \varepsilon \leqslant \sum_{j=1}^{\infty} \sum_{C \in \mathscr{C}(j)} (\operatorname{diam} C)^s + \varepsilon \leqslant \sum_{j=1}^{\infty} \mathcal{H}^s(E_j) + 2\varepsilon.$$

Since $\varepsilon > 0$ is arbitrary, it follows that $\mathcal{H}^s(\bigcup_{j=1}^\infty E_j) \leqslant \sum_{j=1}^\infty \mathcal{H}^s(E_j)$ and the conditions for subadditivity are satisfied. □

In fact $\mathcal{H}^s$ is a metric outer measure, since sets E, F with disjoint closures are a positive distance η say apart. Hence for positive $\delta < \eta$, no element in a δ-cover of E can meet F and vice versa. It follows that for sufficiently small $\delta > 0$,

$$\mathcal{H}^s_\delta(E \cup F) = \mathcal{H}^s_\delta(E) + \mathcal{H}^s_\delta(F),$$

whence

$$\mathcal{H}^s(E \cup F) = \mathcal{H}^s(E) + \mathcal{H}^s(F).$$

Moreover $\mathcal{H}^s$ is a regular outer measure since given any subset A of $\mathbb{R}^n$, there exists a G_δ set E with $\mathcal{H}^s(E) = \mathcal{H}^s(A)$ (see [100, Theorem 1.6]). Usually $\mathcal{H}^s(E)$ is called the *Hausdorff s-dimensional outer measure* of E (but see [189], [103]). The restriction of $\mathcal{H}^s$ to the σ-field of $\mathcal{H}^s$-measurable sets (which includes the open and closed sets and hence G_δ and F_σ sets) is called the *Hausdorff s-dimensional measure*. We will be concerned only with $\mathcal{H}^s$-measurable sets since the number theoretic sets considered are lim-sup or lim-inf sets and reference to outer measure will usually be omitted.

Given a set E, the Hausdorff measure $\mathcal{H}^s(E)$ is non-increasing as s increases. For suppose $0 \leqslant s \leqslant t$ and consider the δ-cover $\mathscr{C}_\delta$ where $\delta < 1$. Then for each $C \in \mathscr{C}_\delta$, $(\operatorname{diam} C)^s \geqslant (\operatorname{diam} C)^t$ whence $\mathcal{H}^s(E) \geqslant \mathcal{H}^t(E)$. The diameter of a set is the same as that of its closure or convex hull and it follows that Hausdorff s-measure will not be affected if each set in the cover is replaced by its closure or its convex hull, or if the covers are restricted to closed sets or to open sets.

The diameter of a set is unchanged by translation, rotation or reflection and it follows that the Hausdorff s-measure is likewise unchanged. However the Hausdorff measure of a set E is altered by a similarity transform T say, as given $x, x' \in \mathbb{R}^n$, $|T(x) - T(x')| = \lambda|x - x'|$ for some $\lambda > 0$. Hence for each C in the cover of E, $(\operatorname{diam} T(C))^s = \lambda^s (\operatorname{diam} C)^s$ and it follows from (3.1) and (3.3) that

$$\mathcal{H}^s(T(E)) = \lambda^s \mathcal{H}^s(E). \tag{3.5}$$

3.2.1. The values assumed by Hausdorff s-measure. It follows readily from the above observations concerning the diameter of a set that to obtain the Hausdorff measure of a set in $\mathbb{R}$, the cover can be taken to consist of intervals. Thus on the real line Hausdorff 1-dimensional measure $\mathcal{H}^1$ coincides with 1-dimensional Lebesgue measure. More generally when n is an integer, Hausdorff n-dimensional measure in $\mathbb{R}^n$ is comparable to n-dimensional Lebesgue measure (two measures

μ, μ' are comparable if for any measurable set E, $\mu(E) \asymp \mu'(E)$) and for each measurable $E \subseteq \mathbb{R}^n$,

$$|E|_{\mathbb{R}^n} = c_n \mathcal{H}^n(E), \tag{3.6}$$

where c_n is the volume of a n-dimensional ball of diameter 1 (see [100, Theorem 1.12], [103, 2.10.2], [189, p. 54, Theorem 30]). It follows that a set of positive n-dimensional Lebesgue measure has positive n-dimensional Hausdorff measure.

Zero dimensional Hausdorff measure $\mathcal{H}^0(E)$ is (as is readily verified) counting measure and so is the cardinality of E. Thus $\mathcal{H}^0(E) = \infty$ when E is an infinite set; for example $\mathcal{H}^0(\mathbb{Q}) = \infty$. For $s > 0$ and each $N = 1, 2, \ldots$, the collection $\{(p/q - 2^{-|p|-q}, p/q + 2^{-|p|-q}) \colon p \in \mathbb{Z}, q \geqslant N\}$, is a cover of intervals of length $2^{1-|p|-q} \leqslant 2^{1-N}$, for $\mathbb{Q}$ with s-length

$$\sum_{q=N}^{\infty} \sum_{p \in \mathbb{Z}} \left(\operatorname{diam} \left(\frac{p}{q} - \frac{1}{2^{|p|+q}}, \frac{p}{q} + \frac{1}{2^{|p|+q}} \right) \right)^s = 2^s \sum_{q=N}^{\infty} \sum_{p \in \mathbb{Z}} \frac{1}{2^{(|p|+q)s}} \leqslant \frac{2^{(3-N)s+1}}{(2^s - 1)^2}.$$

Hence $\mathcal{H}^s_{2^{-N+1}}(\mathbb{Q}) \ll 2^{-Ns}$ and it follows that $\mathcal{H}^s(\mathbb{Q}) = 0$. This behaviour is typical of Hausdorff s-measure and is a consequence of its monotonicity with respect to s.

LEMMA 3.2. *If $\mathcal{H}^s(E) < \infty$ for some $s \geqslant 0$, then for each $\varepsilon > 0$, $\mathcal{H}^{s+\varepsilon}(E) = 0$, while if $\mathcal{H}^s(E) > 0$ for some $s > 0$, then for each $\varepsilon \in (0, s]$, $\mathcal{H}^{s-\varepsilon}(E) = \infty$.*

PROOF. Suppose $\mathcal{H}^s(E) < \infty$. Then by the definition of $\mathcal{H}^s_\delta(E)$, it is possible to choose a δ-cover of E such that

$$\sum_C (\operatorname{diam} C)^s \leqslant \mathcal{H}^s_\delta(E) + 1 \leqslant \mathcal{H}^s(E) + 1 < \infty.$$

Moreover since $\{C\}$ is a δ-cover, $(\operatorname{diam} C)^{s+\varepsilon} \leqslant \delta^\varepsilon (\operatorname{diam} C)^s$, and so

$$\sum_C (\operatorname{diam} C)^{s+\varepsilon} \leqslant \delta^\varepsilon \sum_C (\operatorname{diam} C)^s.$$

Hence by definition,

$$\mathcal{H}^{s+\varepsilon}_\delta(E) \leqslant \sum_C (\operatorname{diam} C)^{s+\varepsilon} \leqslant \delta^\varepsilon \sum_C (\operatorname{diam} C)^s \leqslant \delta^\varepsilon (\mathcal{H}^s(E) + 1).$$

Since $\varepsilon > 0$ and $\mathcal{H}^s(E) < \infty$, it follows that

$$0 \leqslant \mathcal{H}^{s+\varepsilon}(E) = \lim_{\delta \to 0} \mathcal{H}^{s+\varepsilon}_\delta(E) \leqslant (\mathcal{H}^s(E) + 1) \lim_{\delta \to 0} \delta^\varepsilon = 0,$$

i.e., $\mathcal{H}^{s+\varepsilon}(E) = 0$.

Next suppose $\mathcal{H}^s(E) > 0$. If $\mathcal{H}^{s-\varepsilon}(E) < \infty$, then $\mathcal{H}^s(E) = 0$, a contradiction, and the result follows. □

We give a simple consequence of the preceding lemma.

COROLLARY 3.3. *For each $n \in \mathbb{N}$*

$$\mathcal{H}^s(\mathbb{R}^n) = \begin{cases} \infty, & 0 \leqslant s \leqslant n, \\ 0, & s > n. \end{cases}$$

PROOF. Euclidean space $\mathbb{R}^n$ can be written as

$$\mathbb{R}^n = \bigcup_{j=1}^{\infty} [-j, j]^n.$$

The Lebesgue measure $(2j)^n$ of each hypercube $[-j, j]^n$ is finite and so its n-dimensional Hausdorff measure is finite. Hence by the lemma when $s > n$, $\mathcal{H}^s([-j, j]^n) = 0$ for each j and so $\mathcal{H}^s(\mathbb{R}^n)$ vanishes by the subadditivity of Hausdorff measure.

Next by (3.6), $\mathcal{H}^n(\mathbb{R}^n)$ is infinite and so when $s < n$, $\mathcal{H}^s(\mathbb{R}^n)$ is also infinite. □

The sudden drop of the Hausdorff measure from infinity leads to the idea of Hausdorff dimension.

3.3. Hausdorff dimension

It follows from Lemma 3.2 that for each infinite set E in n-dimensional Euclidean space, there exists a unique non-negative exponent s_0 such that

$$\mathcal{H}^s(E) = \begin{cases} \infty, & 0 \leqslant s < s_0, \\ 0, & s_0 < s < \infty, \end{cases}$$

as shown in Figure 3.1. In other words, for all but one value of s, $\mathcal{H}^s(E)$ either is infinite or vanishes. The number

$$s_0 = \inf\{s \in [0, \infty) \colon \mathcal{H}^s(E) = 0\} \tag{3.7}$$

is called the *Hausdorff dimension* of the set E and is denoted by $\dim E$. It is an immediate consequence of the definition that if $\mathcal{H}^s(E)$ vanishes, then $\dim E \leqslant s$; and if $\mathcal{H}^s(E)$ is positive, then $\dim E \geqslant s$.

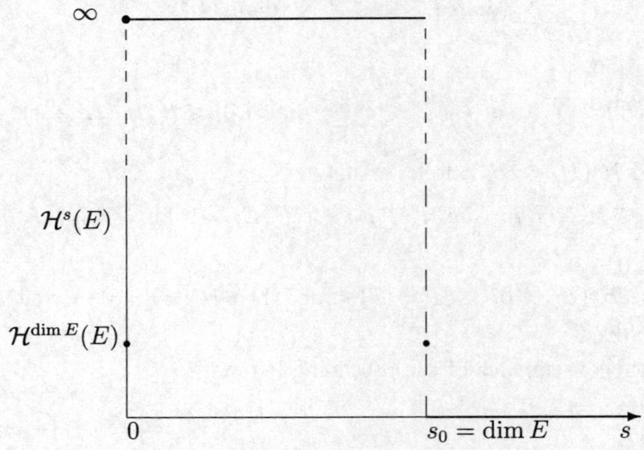

FIGURE 3.1. The graph of the Hausdorff measure of E

The value $\mathcal{H}^{\dim E}(E)$ of the Hausdorff s-measure $\mathcal{H}^s(E)$ when $s = \dim E$ is called the *critical Hausdorff measure* and can be *zero, positive and finite or infinite.* The Hausdorff s-measure of a set consisting of r points is r when $s = 0$ and 0 otherwise, so that the set has zero Hausdorff dimension. Similarly the discussions of the Hausdorff measure of $\mathbb{Q}$ and $\mathbb{R}^n$ above imply that $\dim \mathbb{Q} = 0$ and $\dim \mathbb{R}^n = n$. When I is a (non-empty) interval,

$$\dim I = 1 \tag{3.8}$$

since $\dim I \leqslant 1$ and $\mathcal{H}^1(I) = |I|$, the Lebesgue measure of I.

In the less trivial case of the set $\mathfrak{B}$ of badly approximable numbers, Jarník [125] proved the interesting result that

$$\mathcal{H}^s(\mathfrak{B}) = \begin{cases} \infty, & 0 \leqslant s < 1, \\ 0, & s = 1 \end{cases}$$

($\mathcal{H}^1(\mathfrak{B}) = |\mathfrak{B}| = 0$, see §1.2.3). It follows that $\dim \mathfrak{B} = 1$. He also proved that the set $\mathscr{K}_v$ of v-approximable numbers has Hausdorff dimension $2/(v+1)$ when $v \geqslant 1$ (see §3.5.2 below). Later, in [127], Jarník obtained the Hausdorff measure for the set $\mathscr{S}_v(\mathbb{R}^n)$ of simultaneously v-approximable points and deduced that the critical measure $\mathcal{H}^{2/(v+1)}(\mathscr{K}_v) = \infty$ (see the Notes for more details).

The Cantor 'middle third' set $C(1)$ which, neglecting a countable set, can be regarded as the set of numbers in $[0,1]$ without 1's in its 3-adic expansion is null. It has Hausdorff dimension $\log 2/\log 3$ and critical Hausdorff measure 1 [100, p. 14]. More generally the sets $C(k)$, $k = 0, 1$ or 2, consisting of numbers for which the 3-adic representation lacks the digit k, have Hausdorff dimension $\log 2/\log 3$. However, the critical Hausdorff measure of $C(1)$ is 1 while that of $C(0)$ and $C(2)$ is $0.6457\ldots$ [229]. Thus Hausdorff measure is more discriminating than Hausdorff dimension.

The dependence of the definition of Hausdorff dimension on Hausdorff measure means that the Hausdorff dimension of a set is also unchanged under rigid geometric translations. It is also clear from its definition that the Hausdorff dimension is unchanged under a similarity transform as the critical value of the measure remains the same. These are special cases of the more general result that Hausdorff dimension is invariant under bi-Lipschitz bijections (see Theorem 3.7 below). It makes no difference to the Hausdorff dimension of a set if the dimension is calculated using convex δ-covers.

Hausdorff measure and dimension can be defined for the field of complex numbers $\mathbb{C}$ by regarding $\mathbb{C}$ as $\mathbb{R}^2$; or more generally for $\mathbb{C}^n$ by regarding it as $\mathbb{R}^{2n}$.

3.3.1. Hypercube covers and Hausdorff dimension. It is clear that any measure comparable to Hausdorff measure will have the same critical exponent. Determining the Hausdorff measure and Hausdorff dimension in higher dimensions can be simplified by working with simpler covers consisting of balls or hypercubes. The measures arising from these restricted covers and the Hausdorff measure above are comparable. *Standard hypercube* covers consisting of hypercubes

$$H = \{x \in \mathbb{R}^n\colon |x-a| < \rho\}$$

centred at $a \in \mathbb{R}^n$ and with sides of length 2ρ will be used extensively in Chapters 4 and 5. Because of this, an outer measure $\mathscr{H}^s$ based on covers restricted to standard hypercubes is defined and then shown to be comparable to Hausdorff measure. It is comparable to *spherical* measure (though disc or ball measure might be better terminology), where the covers are restricted to balls [157, Chapter 5].

Let $\mathscr{C}$ be a cover of E by standard hypercubes H of diameter at most δ. The length of a side of the hypercube H will be denoted by $\ell(H)$, so that $\operatorname{diam} H = n^{1/2}\ell(H)$. For any set E in $\mathbb{R}^n$ and $\delta > 0$, define

$$\mathscr{H}^s_\delta(E) = \inf\{\sum_H \ell(H)^s\}$$

where the infimum is over all countable covers of E by standard hypercubes of sidelength $\ell(H) \leqslant \delta$. Then proceeding as in the case $\mathcal{H}^s_\delta$, it can be readily shown that $\mathscr{H}^s_\delta$ and $\mathscr{H}^s = \lim_{\delta\to 0}\mathscr{H}^s_\delta$ are outer measures. Moreover $\mathscr{H}^s$ is comparable to $\mathcal{H}^s$.

LEMMA 3.4. *For each subset E of $\mathbb{R}^n$ and each real number $s \geqslant 0$,*

$$n^{-s/2}\mathcal{H}^s(E) \leqslant \mathscr{H}^s(E) \leqslant \mathcal{H}^s(E).$$

PROOF. Let $\varepsilon > 0$ be an arbitrary positive number and let $\{H\}$ be a cover of E by standard hypercubes with $\ell(H) \leqslant \delta$ and with $\sum_H \ell(H)^s \leqslant \mathscr{H}^s_\delta(E) + \varepsilon$. Then

$$\sum_H \ell(H)^s = n^{-s/2}\sum_H (\operatorname{diam} H)^s = n^{-s/2}\ell^s(\{H\}),$$

and so by the definition of $\mathcal{H}^s_\delta$ and $\mathscr{H}^s_\delta$,

$$\mathcal{H}^s_{\delta n^{-s/2}}(E) \leqslant \sum_H (\operatorname{diam} H)^s \leqslant n^{s/2}\sum_H \ell(H)^s \leqslant n^{s/2}\left(\mathscr{H}^s_\delta(E) + \varepsilon\right),$$

whence $\mathcal{H}^s(E) \leqslant n^{s/2}\mathscr{H}^s(E)$ on letting $\delta \to 0$.

On the other hand, take $\{C\}$ to be a δ-cover of E. Each C is contained in a standard hypercube H with $\operatorname{diam}(C) = \ell(H)$. Hence $\{H\}$ is a cover of E by hypercubes with $\ell(H) < \delta$ and by definition,

$$\mathscr{H}^s_\delta(E) \leqslant \sum_H \ell(H)^s = \sum_C (\operatorname{diam} C)^s.$$

Taking the infimum over δ-covers $\{C\}$ for E gives $\mathscr{H}^s_\delta(E) \leqslant \mathcal{H}^s_\delta(E)$ and the lemma follows on letting $\delta \to 0$. □

It is worth repeating that since $\mathscr{H}^s$ is comparable to $\mathcal{H}^s$, $s_0 = \dim E$ is also the critical exponent for $\mathscr{H}^s(E)$. Standard hypercube covers will be used a great deal in the subsequent analysis of Hausdorff dimension. Net measures, in which the covers are made up of dyadic cubes, are also comparable to Hausdorff measure. They have some similarities with the Cantor construction and have a number of technical advantages. More details are given in [100], [157], [189].

3.4. Properties of Hausdorff dimension

The Hausdorff dimension of a set enjoys many but not all properties of the more familiar kinds of dimension, in particular by (3.6) it coincides with the usual kinds of dimension when the critical exponent s_0 is an integer.

THEOREM 3.5. *Let* E, F *be subsets of* $\mathbb{R}^n$.

(a) *If* $E \subseteq F$ *then* $\dim E \leqslant \dim F$.
(b) $\dim E \leqslant n$.
(c) *When* $|E| > 0$, *then* $\dim E = n$.
(d) *The dimension of a point is* 0.
(e) *If* $\dim E < n$, *then* $|E| = 0$.

Each part is a straightforward consequence of the properties of Hausdorff measure.

(a) follows from the monotonicity of Hausdorff outer s-measure.
(b) holds since $\dim E \leqslant \dim \mathbb{R}^n = n$ for any $E \subseteq \mathbb{R}^n$.
(c) for n-dimensional Hausdorff measure and Lebesgue measure are comparable, whence $\mathcal{H}^n(E) > 0$ and so $\dim E \geqslant n$. This and (b) give the result.
(d) holds since a point can be covered by a ball of arbitrarily small radius.
(e) follows from (b) and the contrapositive of (c).

Another useful property, sometimes called countable stability [100], is now proved to illustrate the common technique of establishing the upper and lower bounds for the Hausdorff dimension separately.

LEMMA 3.6. *Let* $E = \bigcup_{j=1}^{\infty} E_j$. *Then*

$$\dim E = \sup\{\dim E_j \colon j = 1, 2, \dots\}.$$

PROOF. For each $j = 1, 2, \dots$, the set $E_j \subseteq E$, whence by monotonicity, $\dim E_j \leqslant \dim E$ and

$$\sup\{\dim E_j \colon j = 1, 2, \dots\} \leqslant \dim E.$$

To prove the complementary inequality, observe that if $\sup\{\dim E_j \colon j = 1, 2, \dots\}$ were infinite, then $\dim E = \infty$. Otherwise let $s = \sup\{\dim E_j \colon j = 1, 2, \dots\} + \varepsilon$, where $\varepsilon > 0$ is arbitrary. Then for each j, $\mathcal{H}^s(E_j) = 0$. But by subadditivity, $\mathcal{H}^s(E) = \mathcal{H}^s(\bigcup_j E_j) \leqslant \sum_j \mathcal{H}^s(E_j) = 0$, whence $s = \sup \dim E_j + \varepsilon \geqslant \dim E$ and as ε is arbitrary $\sup \dim E_j \geqslant \dim E$. □

This lemma allows one (as with Lebesgue measure) to replace the study of the Hausdorff dimension of sets in $\mathbb{R}^n$ with sets in I^n for a suitable interval I. It also implies that the Hausdorff dimension of a countable set is 0 (see Lemma 3.6 above) and that the dimensions of sets differing only by a countable set are equal. There are, however, uncountable sets of zero Hausdorff dimension (for example the set of Liouville numbers, see §3.5.3 below) and there are null sets with maximal Hausdorff dimension(for example the set of badly approximable numbers $\mathfrak{B}$ is null but $\dim \mathfrak{B} = 1$ [125], as mentioned above in §3.3).

The Hausdorff dimension fails to behave like a dimension for Cartesian products since only the downward inequality

$$\dim(A \times B) \geqslant \dim A + \dim B$$

holds in general (see [100], [157]). However this inequality evidently provides a lower bound and equality holds if one of the sets is sufficiently nice, for example if its Hausdorff dimension coincides with its upper packing dimension [157]. In particular equality holds for cylinder sets $A \times I$, where I is an interval (see §3.5.5 below) and can be useful in practice (see for example Chapter 7 and [53], [73], [81], [85]).

3.4.1. Bi-Lipschitz invariance. The Hausdorff dimension of a set is preserved under a bi-Lipschitz bijection (see also [100, Lemma 1.8]). As a result, the Hausdorff dimension of subsets of a manifold can be deduced from that of their image under a chart in Euclidean parameter space.

THEOREM 3.7. *Let $f\colon E \to F$ be a function between sets in Euclidean space.*
(a) If f is Lipschitz on E, then $\dim f(E) \leqslant \dim E$.
(b) If f is a bi-Lipschitz bijection on E, then $\dim E = \dim f(E)$.

PROOF. (a) We can suppose without loss of generality that $\dim E$ is finite. Let $s = \dim E + \eta$, where $\eta > 0$ is arbitrary. Then given $\varepsilon > 0$, there exists a δ-cover $\{C\}$ of E such that

$$\sum_C (\operatorname{diam} C)^s < \varepsilon.$$

But $\{f(C)\}$ is a cover of $f(E)$ with

$$\begin{aligned}\operatorname{diam} f(C) &= \sup\{|f(x) - f(y)|\colon x, y \in C\}\\ &\leqslant c \sup\{|x - y|\colon x, y \in C\} = c \operatorname{diam} C,\end{aligned}$$

where $c > 0$ is the Lipschitz constant for f. Hence the sum $\sum_C (\operatorname{diam} f(C))^s < c^s \varepsilon$. It follows that $\mathcal{H}^s(f(E)) = 0$ and so $\dim f(E) \leqslant \dim E + \eta$. But η is arbitrary, whence the result.

(b) This follows by considering the Lipschitz inverse function of f and using (a). □

This theorem is used constantly in analysing the metric structure of manifolds, as will be seen in the following sections. The following simple application is useful (see §4.2.3, [15] and [34]).

COROLLARY 3.8. *Let $S \subseteq \mathbb{R} \setminus \{0\}$ and let $S^{-1} = \{s^{-1}\colon s \in S\}$. Then*

$$\dim S^{-1} = \dim S.$$

PROOF. Express $\mathbb{R} \setminus \{0\}$ as a countable union of closed bounded subintervals I:

$$\mathbb{R} \setminus \{0\} = \bigcup_{n=1}^{\infty} \left(\left[\frac{1}{n+1}, \frac{1}{n}\right] \cup [n, n+1] \cup \left[-\frac{1}{n}, -\frac{1}{n+1}\right] \cup [-n-1, -n] \right).$$

Then $S = \bigcup_I (S \cap I)$ and $S^{-1} = \bigcup_I (S \cap I)^{-1}$. The function $f\colon \mathbb{R} \setminus \{0\} \to \mathbb{R} \setminus \{0\}$ given by $f(x) = 1/x$ is bi-Lipschitz on each interval I, since for each interval I of positive numbers,

$$(\sup I)^{-2}|x - x'| \leqslant |f(x) - f(x')| = \frac{|x - x'|}{xx'} \leqslant (\inf I)^{-2}|x - x'|$$

and similarly for any interval I of negative numbers. Thus f is bi-Lipschitz on each $S \cap I$ and therefore $\dim(S \cap I) = \dim(S \cap I)^{-1}$. But

$$\dim S^{-1} = \sup_I \left(\dim (S \cap I)^{-1}\right) = \sup_I \dim(S \cap I) = \dim S,$$

as claimed. □

3.5. Determining the Hausdorff dimension

Unless some general result is available, the determination of the Hausdorff dimension $\dim E$ of a set E commonly falls into two parts, with the upward inequality $\dim E \leqslant d$ say and the complementary downward inequality $\dim E \geqslant d$ being established separately. A simple example is given by the proof of Lemma 3.6. It is worth repeating that determining Hausdorff dimension is often simpler using covers consisting of hypercubes H and the corresponding measure $\mathscr{H}^s$. Since $\ell(H) = n^{-1/2} \operatorname{diam} H$, it is immaterial whether the diameter or sidelength is used in computing the s-length.

3.5.1. Upper bounds. Suppose E is a null set. A value of s for which $\mathcal{H}^s(E)$ vanishes is an upper bound for the Hausdorff dimension of E. To find such a value, it suffices to exhibit a *hypercube* cover $\{C\}$ of E with hypercubes of arbitrarily small sidelength and s-length. It is sometimes possible to do this simply by changing the exponent n in the cover used to obtain the Lebesgue measure of E to s; see for example [58], [145], [210].

LEMMA 3.9. *Let $\varepsilon > 0$ be given. Suppose that there exists a $\delta\,(= \delta(\varepsilon)) > 0$ such that for each positive $\rho < \delta$, there exists a ρ-cover $\mathscr{C}$ for E with s-length satisfying*

$$\ell^s(\mathscr{C}) = \sum_{C \in \mathscr{C}} (\operatorname{diam} C)^s < \varepsilon.$$

Then $\mathcal{H}^s(E) = 0$ and $\dim E \leqslant s$.

PROOF. By definition, $\mathcal{H}^s_\rho(E) \leqslant \sum_{C \in \mathscr{C}} (\operatorname{diam} C)^s < \varepsilon$ and for ρ sufficiently small,

$$\mathcal{H}^s(E) \leqslant \mathcal{H}^s_\rho(E) + \varepsilon \leqslant 2\varepsilon,$$

so that $\mathcal{H}^s(E) = 0$ and $\dim E \leqslant s$. □

The following Hausdorff measure counterpart of Cantelli's lemma for lim-sup sets is simple but useful. It will be used repeatedly and will be referred to as the Hausdorff-Cantelli lemma.

LEMMA 3.10 (HAUSDORFF-CANTELLI). *Let E be a set in $\mathbb{R}^n$ and suppose that*

$$E \subseteq \{t \in \mathbb{R}^n : t \in H_j \text{ for infinitely many } j \in \mathbb{N}\}$$

where $H_j, j \in \mathbb{N}$, is a family of hypercubes. If for some $s > 0$,

$$\sum_{j=1}^{\infty} \ell(H_j)^s < \infty,$$

then $\mathscr{H}^s(E) = 0$ and $\dim E \leqslant s$.

PROOF. From the hypotheses,

$$E \subseteq \bigcap_{N=1}^{\infty} \bigcup_{j=N}^{\infty} H_j.$$

Thus for each $N = 1, 2, \ldots$, the family $\mathscr{C}_N = \{H_j : j \geqslant N\}$ is a cover for E such that given $\rho > 0$, $\ell(H_j) < \rho$ for each $j \geqslant N_1 = N_1(\rho)$ since $\sum_{j=1}^{\infty} \ell(H_j)^s < \infty$. Moreover given any $\varepsilon > 0$

$$\sum_{j=N}^{\infty} \ell(H_j)^s < \varepsilon$$

for N sufficiently large. Hence $\mathscr{H}_\rho^s(E) < \varepsilon$ and the lemma follows on letting $\rho \to 0$. □

When the natural cover for a lim-sup set can be refined to a hypercube cover, an upper estimate for the Hausdorff dimension can be established using the Hausdorff-Cantelli lemma. This approach is very effective in many cases when the variables are independent.

For instance to show that $\dim (A \times I) \leqslant \dim A + \dim I$, where I is a finite interval, consider a cover of A by small hypercubes H with $\sum_H \ell(H)^s < \infty$, where $s > \dim A$. Cover $A \times I$ by hypercubes $H \times I'$, where I' is a subinterval of I of length $\ell(H)$. The collection $\{H \times I'\}$ of hypercubes covers $A \times I$ and

$$\sum_{H \times I'} \ell(H \times I')^{s+1} \ll \sum_{H} \ell(H)^{s+1} \frac{\ell(I)}{\ell(H)} \ll \sum_{H} \ell(H)^s < \infty,$$

whence

$$\dim (A \times I) \leqslant \dim A + 1 \tag{3.9}$$

(by part (c) of Theorem 3.5, $\dim I = 1$). The complementary inequality will be dealt with in the next chapter.

3.5.2. The Jarník-Besicovitch theorem. As another illustration of the use of natural covers, we obtain the correct upper bound for Hausdorff dimension of $\mathscr{K}_v$, the set of v-approximable numbers (defined in §1.2.1). The simple approach can be inefficient for subsets of manifolds when the variables are dependent and resonant sets which are near to tangential give rise to substantial volume contributions in the relevant sum. Getting round this difficulty will constitute a great part of Chapter 4. The Hausdorff dimension of $\mathscr{K}_v$ was determined by Jarník [126] and independently by A. S. Besicovitch [48] later (see also [54], [98] and the Notes).

THEOREM 3.11 (JARNÍK-BESICOVITCH). *When* $v \geqslant 1$, $\dim \mathscr{K}_v = 2/(v+1)$.

Of course when $v \leqslant 1$, $\mathscr{K}_v = \mathbb{R}$ (see Chapter 1). The easy part of this theorem will be proved by obtaining the correct upper bound for the Hausdorff dimension. There is no loss of generality in restricting to $[0,1]$ and considering

$$\mathscr{K}_v = \{\xi \in [0,1] \colon \|q\xi\| < q^{-v} \text{ for infinitely many } q \in \mathbb{N}\} = \limsup_{q\to\infty} B(q),$$

where $v > 1$ and $B(q) = \{\xi \in [0,1] \colon \|q\xi\| < q^{-v}\}$. The family of sets $B(q)$, $q \in \mathbb{N}$, is evidently a natural cover for $\mathscr{K}_v$ and

$$B(q) = \bigcup_{p=0}^{q} \left(\frac{p}{q} - \frac{1}{q^{v+1}}, \frac{p}{q} + \frac{1}{q^{v+1}} \right) \cap [0,1],$$

a union of intervals, each of length at most $2/q^{v+1}$. Thus these intervals are a natural interval cover for $\mathscr{K}_v$ and each $\xi \in \mathscr{K}_v$ lies in infinitely many intervals $(p/q - 1/q^{v+1}, p/q + 1/q^{v+1})$. The s-length $\ell^s(\mathcal{C})$ of this natural interval cover $\mathcal{C}$ satisfies

$$\ell^s(\mathcal{C}) \leqslant \sum_{q=1}^{\infty} \sum_{p=0}^{q} 2^s q^{-(v+1)s} \ll \sum_{q=1}^{\infty} q^{1-(v+1)s} < \infty$$

when $s > 2/(v+1)$. Hence by the Hausdorff-Cantelli lemma

$$\dim \mathscr{K}_v \leqslant 2/(v+1) \tag{3.10}$$

when $v > 1$. The more difficult complementary inequality is discussed in Chapter 5.

3.5.3. Liouville numbers. A real number α is a Liouville number if for any $N > 0$, there exist infinitely many rationals p/q such that

$$|\alpha - p/q| < q^{-N}.$$

It follows from (3.10) that the Hausdorff dimension of the set of Liouville numbers is 0, since it is less than $2/(1+N)$ for any N.

3.5.4. Lower bounds. Establishing the complementary downward inequality $\dim E \geqslant d$ for the Hausdorff dimension of the set E entails showing that given *any* $s < d$ and *any* cover $\{C\}$ of E with the diameters of the covering elements arbitrarily small, the s-length $\sum_C (\operatorname{diam} C)^s \geqslant \delta$ for some positive δ. This can be very difficult and often requires resorting to a variety of methods, for example part (b) of Theorem 3.5 (see [73], [96]). The regular systems introduced by A. Baker and W. M. Schmidt [13] and the related ubiquitous systems [90] have proved effective in obtaining lower bounds for the Hausdorff dimension of sets of number theoretic interest. This approach depends on the approximating elements (the rationals in the case of $\mathscr{K}_v$) having a 'regular' distribution in the sense that enough of them are reasonably well separated. The more fundamental mass distribution principle is discussed in §3.5.7 below and will be applied in Chapter 5.

3.5.5. The Hausdorff dimension of cylinder sets. Net measures can be used to show that $\dim(A \times B) \geqslant \dim A + \dim B$ for sets $A \subseteq \mathbb{R}^m$ and $B \subseteq \mathbb{R}^n$ [100, 157] but there is a simple proof for the cylinder set $A \times I$ where I is a finite interval (see [53]). Let $\dim A = d$ and suppose that $\dim(A \times I) < d + 1$. Then by definition, there exists a cover $\mathscr{C} = \{H_j \colon j \in \mathbb{N}\}$ of $A \times I$ by hypercubes H_j with

$$\ell^{d+1-\delta}(\mathscr{C}) \ll \sum_{j=1}^{\infty} \ell(H_j)^{d+1-\delta} < \varepsilon < \infty \tag{3.11}$$

for some $\delta > 0$. For each $\alpha \in I$, there is a cover $\mathscr{C}(\alpha)$ consisting of n-dimensional hypercubes $\widehat{H}_j(\alpha) = H_j \cap (\mathbb{R}^m \times \{\alpha\})$ say where $\ell(\widehat{H}_j(\alpha)) = \ell(H_j)$ of the cross-section $A \times \{\alpha\}$. Define $\chi_j \colon I \to \mathbb{R}$ by

$$\chi_j(\alpha) = \begin{cases} 1, & \text{if } H_j \cap (\mathbb{R}^n \times \{\alpha\}) \neq \emptyset, \\ 0, & \text{otherwise.} \end{cases}$$

Then $\chi_j(\alpha) \geqslant 0$, $\int_I \chi_j(\alpha)\,d\alpha = \ell(H_j)$ and $\sum_j \ell(\widehat{H}_j(\alpha))^{d-\delta} = \sum_{j=1}^{\infty} \chi_j(\alpha)\,\ell(H_j)^{d-\delta}$, whence

$$\sum_j \ell(H_j)^{d+1-\delta} = \sum_{j=1}^{\infty} \ell(H_j)^{d-\delta} \left(\int_I \chi_j(\alpha)\,d\alpha \right) = \int_I \left(\sum_{j=1}^{\infty} \chi_j(\alpha)\,\ell(H_j)^{d-\delta} \right) d\alpha < \varepsilon$$

by the monotone convergence theorem and (3.11). Thus there exists an $\alpha \in I$ such that

$$\ell^{d-\delta}(\mathscr{C}(\alpha)) \ll \sum_{j=1}^{\infty} \ell(\widehat{H}_j(\alpha))^{d-\delta} = \sum_{j=1}^{\infty} \chi_j(\alpha)\,\ell(H_j)^{d-\delta} < \varepsilon/|I|$$

and evidently $A \times \{\alpha\}$ has a cover $\mathscr{C}(\alpha)$ with $(d-\delta)$-length $\ll \varepsilon/|I|$. It follows from the definition that $\dim(A \times \{\alpha\}) < d$. But the map $A \to A \times \{\alpha\}$ given by $a \mapsto (a, \alpha)$ is a bi-Lipschitz bijection and hence $\dim A = \dim(A \times \{\alpha\}) < d$, a contradiction. In view of (3.9),

$$\dim(A \times I) = \dim A + 1. \tag{3.12}$$

3.5.6. The exact exponent of approximation. The Hausdorff dimension of the set of points in $\mathbb{R}^n$ approximable to exact exponent v is clearly at most that of the set of points approximable to the exponent v. If, however, the Hausdorff measure of the latter set at the critical exponent is positive, the two sets have the same Hausdorff dimension. This was proved by R. Güting [113] for the set

$$\mathcal{K}'_v = \{\xi \in \mathbb{R} \colon \omega(\xi) = v\},$$

where $\omega(\xi) = \sup\{w \colon \xi \in \mathscr{K}_w(\mathbb{R})\}$, of real numbers approximable by rationals to exact exponent v (or exact order $v+1$), introduced in §1.3.3, and we give a proof. Let $v \geqslant 1$. The upward inequality $\dim \mathcal{K}'_v \leqslant 2/(v+1)$ is immediate as $\mathcal{K}'_v \subset \mathcal{K}_v$ and since $\dim \mathcal{K}_v = \dim \mathscr{K}_v$ (see §3.5.2 above). The complementary inequality follows from representing $\mathcal{K}_v$ as

$$\mathcal{K}_v = \mathcal{K}'_v \cup \left(\bigcup_{r=1}^{\infty} \mathcal{K}_{v+\frac{1}{r}} \right)$$

and then observing for each $r \geqslant 1$, $\mathcal{H}^{2/(v+1)}(\mathcal{K}_{v+1/r}) = 0$, so that $\mathcal{H}^{2/(v+1)}(\mathcal{K}'_v) = \mathcal{H}^{2/(v+1)}(\mathcal{K}_v)$. But Jarník showed in [127] that $\mathcal{H}^{2/(v+1)}(\mathscr{K}_v)$ is infinite, whence $\mathcal{H}^{2/(v+1)}(\mathcal{K}_v)$ is infinite and $\mathcal{H}^{2/(v+1)}(\mathcal{K}'_v)$ likewise. Thus $\dim \mathcal{K}'_v \geqslant 2/(v+1)$ and equality follows.

This kind of argument was invoked by A. Baker and W. M. Schmidt in [13] to deduce that $\dim \mathfrak{K}'_w = \dim \mathfrak{K}_w$ (defined below in §4.1.2) and used by R. C. Baker in [16] in his study of planar curves (see §5.2.4 below).

3.5.7. Mass distributions and capacity. Hausdorff measure is closely related to the notion of capacity which has its origins in electrostatics and potential theory. A measure which is supported on a bounded subset of $\mathbb{R}^n$ and which satisfies $0 < \mu(\mathbb{R}^n) < \infty$ is called a *mass* or *charge distribution*. For each $s > 0$, the s-capacity $\mathrm{C}_s(E)$ of a set E is defined to be the supremum of $1/I_s(\mu)$ over all mass distributions μ with $\mu(E) = 1$, where

$$I_s(\mu) = \int_E \int_E \frac{d\mu(x) d\mu(y)}{|x-y|_2^s}. \tag{3.13}$$

The s-capacity is a measure but can be far from additive (*e.g.,* when $s \leqslant n-2$, the s-capacity $\mathrm{C}_s(B^n(a,r))$ of the open ball in $\mathbb{R}^n$ centred at a and of radius r is the same as that of the closure and the boundary $S^n(a,r)$ [157, p. 111]).

Capacity gives rise to the capacity dimension $\dim_C$, where

$$\dim_C(E) = \inf\{s \colon \mathrm{C}_s(E) = 0\} = \sup\{s \colon \mathrm{C}_s(E) > 0\},$$

which is similar to and indeed for Borel sets coincides with Hausdorff dimension [157, Theorem 8.9]. The capacity dimension can give a lower bound for the

Hausdorff dimension of a set. Frostman's lemma is important in this connection and is discussed briefly below (more details are in [157, §8.4]; a full treatment with alternative definitions is in [117]).

Let μ be a mass distribution on a subset E of $\mathbb{R}^n$. Suppose that for some $s \geqslant 0$ there are positive constants c and δ such that $\mu(H) \leqslant c\,\ell(H)^s$ for any hypercube H in $\mathbb{R}^n$ with sidelength $\ell(H) \leqslant \delta$. Then for any set E in $\mathbb{R}^n$, $\mathscr{H}^s(E) \geqslant \mu(E)/c$. The proof is short. Let $\{H_i\}$ be a δ-cover of E by hypercubes H_i. Then

$$0 < \mu(E) \leqslant \mu\left(\bigcup_i H_i\right) \leqslant \sum_i \mu(H_i) \leqslant c \sum_i \ell(H)^s.$$

Taking infima over all such covers, we see that $\mathscr{H}^s_\delta(E) \geqslant \mu(E)/c$, whence on letting $\delta \to 0$,

$$\mathscr{H}^s(E) \geqslant \mu(E)/c. \tag{3.14}$$

Since a ball can always be fitted between two hypercubes of comparable radius, this gives the easy part of Frostman's lemma which is now stated.

LEMMA 3.12. *Let E be a Borel subset of $\mathbb{R}^n$. Then*

$$\mathcal{H}^s(E) > 0$$

if and only if E carries a probability measure μ such that for some positive constant c, $\mu(B) \leqslant c\,(\operatorname{diam} B)^s$ for all balls B.

COROLLARY 3.13. *If E supports a probability measure μ with $\mu(B) \ll (\operatorname{diam} B)^s$ for all sufficiently small balls B, then* $\dim E \leqslant s$.

For if $\operatorname{diam} B \geqslant c$ for some constant $c > 0$, then

$$\mu(B) \leqslant \mu(E) c^s \operatorname{diam}(B) \ell^s \ll \ell(B)^s.$$

This is often called the *mass distribution principle*; the more difficult converse can be proved using net measures (see [57], [101], [157]).

In the complex plane $\mathbb{C}$, the logarithmic capacity $\mathrm{C}_0(E)$ is defined as in (3.13) but with denominator $-\log|x-y|_2$. It can also be defined geometrically for compact sets E by

$$\mathrm{C}_0(E) = \lim_{r\to\infty} \Delta_r\,(E)^{\frac{1}{r(r-1)}},$$

where

$$\Delta_r\,(E) = \max_{z_1,\ldots,z_r \in E} \prod_{i\neq j} |z_i - z_j|^2, \; r = 2, 3, \ldots,$$

is called the r-discriminant [184, Chapter 11] ($\mathrm{C}_0(E)$ is also called the transfinite diameter of E).

It follows readily from the definition that as well as being monotonic and invariant under translation, capacity is homogeneous (when $f(z) = az$, $\mathrm{C}_0(f(E)) = |a|\,\mathrm{C}_0(E)$) and enjoys a contraction property: if $|\varphi(z) - \varphi(z')| \leqslant |z - z'|$ for $z, z' \in E$, then $\mathrm{C}_0(\varphi(E)) \leqslant \mathrm{C}_0(E)$. These properties enable capacity to be used to study integral polynomials and to estimate the Hausdorff dimension of certain number theoretic sets (see §4.1.3).

3.6. Hausdorff dimension on manifolds

The decomposition of a smooth m-dimensional manifold M in $\mathbb{R}^n$ into at most countably many submanifolds (see §1.4.1) which provides a natural framework for the study of Lebesgue measure on M, also serves for the study of the Hausdorff dimension of relatively null subsets of M. Recall that U can be taken to be a small hypercube and the chart $h\colon M_U \to U$ can be assumed to be bi-Lipschitz on M_U ($h = \pi_U$ when $\theta = 1_U \times \varphi$). By Theorem 3.7, for any set $X \subset M$, $\dim(X \cap M_U) = \dim h(X \cap M_U)$ and so questions about the Hausdorff dimension of subsets of the manifold can be transferred to sets in Euclidean space (just as with Lebesgue measure). Since $\dim U = m$ and $\dim M_U = \dim U$, one immediate consequence from this and Lemma 3.6 is that the Hausdorff dimension of an m-dimensional smooth manifold M is m.

The most important consequence for us concerns the various sets of ψ-approximable points on the manifold. In particular, by Lemma 3.6,

$$\dim \mathscr{S}(M;\psi) = \sup_U \dim \mathscr{S}(M_U;\psi) \tag{3.15}$$

and

$$\dim \mathscr{L}(M;\psi) = \sup_U \dim \mathscr{L}(M_U;\psi), \tag{3.16}$$

so that we can restrict attention to the subsets M_U on M. By Theorem 3.7 and the choice of U,

$$\dim \mathscr{S}(M_U;\psi) = \dim h(\mathscr{S}(M_U;\psi)) = \dim S(\psi) \tag{3.17}$$

and

$$\dim \mathscr{L}(M_U;\psi) = \dim h(\mathscr{L}(M_U;\psi)) = \dim L(\psi), \tag{3.18}$$

where we recall that the dependence of $S(\psi)$ and $L(\psi)$ on the manifold and the reference to the parametrisation are suppressed (see (1.27) and (1.28)). Recall further that it suffices to study $S(\psi)$ and $L(\psi)$ when U is a sufficiently small hypercube in $\mathbb{R}^m$.

It is shown in §1.3.3 that the set $\mathcal{S}_v(M)$ satisfies

$$\mathscr{S}_v(M) \subset \mathcal{S}_v(M) \subset \mathscr{S}_{v-\varepsilon}(M)$$

for any $\varepsilon > 0$. Hence $\dim \mathscr{S}_v(M) \leqslant \dim \mathcal{S}_v(M) \leqslant \dim \mathscr{S}_{v-\varepsilon}(M)$ and similarly $\dim \mathscr{L}_v(M) \leqslant \dim \mathcal{L}_v(M) \leqslant \dim \mathscr{L}_{v-\varepsilon}(M)$. Thus when the Hausdorff dimension of one of the sets is known to depend continuously on v, the other sets have the same Hausdorff dimension.

By extending results of A. Baker and Schmidt [13], R. C. Baker has shown for a smooth planar curve Γ with curvature non-zero except on a set of zero Hausdorff dimension that

$$\dim \mathscr{L}_v(\Gamma) = \dim \mathcal{L}_v(\Gamma) = \dim \mathcal{L}'_v(\Gamma) = 3/(v+1)$$

when $v \geqslant 2$ [16]. When M is a smooth (C^3) manifold in $\mathbb{R}^n$ of dimension $m \geqslant 2$ which is 2-curved except on a set of Hausdorff dimension $m-1$, we have

$$\dim \mathscr{L}_v(M) = \dim \mathcal{L}_v(M) = \dim \mathcal{L}'_v(M) = m - 1 + \frac{n+1}{v+1} \tag{3.19}$$

for $v \geqslant n$ [89]. By [33], the above formula also holds for $M = \mathscr{V}$ (where $m = 1$) and it follows that $\mathfrak{M}_v$, $\mathscr{M}_v$ and $\mathscr{M}'_v$ all have Hausdorff dimension $(n+1)/(v+1)$ for $v \geqslant n$. The Hausdorff dimension of $\mathscr{S}(M;\psi)$ and $\mathscr{L}(M;\psi)$ will be discussed further in Chapters 4 and 5.

3.7. Notes

A historical note In his more general approach [56] of 1914 to Lebesgue's theory of measure, Carathéodory used covers consisting of arbitrary sets with sufficiently small diameters. Although he was mainly concerned with linear measure, in the final part of [56] Carathéodory indicated how a p-dimensional measure in $\mathbb{R}^q$ could be constructed for any non-negative integer $p \leqslant q$. His construction was based on extending to higher dimensions the observation that the diameter of a set in $\mathbb{R}^q$ is the supremum of the diameters of the orthogonal projections of the convex hull of the set onto all possible straight lines in $\mathbb{R}^q$ through the origin. The p-diameter $\operatorname{diam}^{(p)}$ say was defined to be the supremum of the p-dimensional volumes of all orthogonal projections of the convex hull onto all p-dimensional vector subspaces of $\mathbb{R}^q$. Carathéodory's p-dimensional Lebesgue measure of the set E in $\mathbb{R}^q$ is based on the numerical sum

$$\sum_{C \in \{C\}} \operatorname{diam}^{(p)}(C)$$

where $\{C\}$ is a cover of E with $\operatorname{diam}^{(p)}(C)$ small. This prompted Hausdorff's seminal paper of 1919 [116] based on the far-reaching observation (which he modestly described as a 'small contribution' to Carathéodory's work) that Carathéodory's definition of linear measure could be extended in a natural way to another p-dimensional measure. In Hausdorff's extension, the p-dimensional measure of a set E in Euclidean space was based on the sum

$$\sum_{C \in \{C\}} (\operatorname{diam} C)^p.$$

In this formulation, the various volume formulae and estimates still make sense when the integral dimension p is replaced by a real number s. Hausdorff normalised his estimates by a factor $c_s = \pi^{s/2}2^{-s}/(s/2)!$, the volume of the s-dimensional ball of diameter 1. In this way he constructed an s-dimensional outer measure where s is *any* non-negative real number s (instead of as hitherto a non-negative integer) for any set in $\mathbb{R}^n$; and he determined the measure and the dimension of the classical Cantor set. Actually Hausdorff developed the theory with a non-negative summand $l(C)$ satisfying certain natural conditions; Rogers develops Hausdorff measure in terms of a 'dimension' function [189, Chapter 2]. We shall not be concerned with this more general line of development but [189] contains a full account. The particular dimension arising from the s-power function $s \mapsto x^s$ is also called the fractional dimension or Hausdorff-Besicovitch dimension.

In 1912 E. Borel [51] proved that the set $\mathscr{K}_v$ of v-approximable numbers was null for $v > 2$. This was extended by Khintchine, who wrote a series of papers in the 1920's on the Lebesgue measure of the sets $\mathscr{K}(\psi)$ of ψ-approximable and $\mathfrak{B}$ of badly approximable numbers [134, 135, 136, 137]. In his pioneering paper [125] of 1928/9, Jarník proved that $\dim \mathfrak{B} = 1$ and obtained estimates for the Hausdorff measure of sets of numbers with restricted coefficients in their continued fraction expansions (there is a discussion in [189]). At about the same time, he proved that $\dim \mathscr{K}_v = 2/(v+1)$ for $v \geqslant 1$ [126]; the proof is lengthy and involves continued fractions and other number theoretic ideas. Incidentally, Besicovitch's first paper on Hausdorff dimension (on the geometry of planar sets) was also published in 1929 [47]. Besicovitch [48] proved Jarník's result later (in 1934) independently and more simply using geometric ideas and it is now called the Jarník-Besicovitch theorem. Neither obtained the measure at the critical exponent but in 1931, Jarník [127] extended the theorem to a Hausdorff measure version of Khintchine's theorem (Corollary 1.4 (b)) and under certain monotonicity assumptions essentially showed that $\mathscr{H}^s(\mathscr{S}(\mathbb{R}^n;\psi))$ vanishes if the sum $\sum_{r=1}^{\infty} \psi(r)^s r^{n-s}$ converges and is infinite otherwise. This implies that $\dim \mathscr{S}_v(\mathbb{R}^n) = (n+1)/(v+1)$ for $v \geqslant 1/n$ (see also [98]) and that the Hausdorff measure of $\mathscr{S}_v(\mathbb{R}^n)$ is infinite at the critical exponent $(n+1)/(v+1)$. It follows by continuity and from the inclusions (1.15) that the Hausdorff dimensions of $\mathscr{K}_v$ and $\mathcal{K}_v$ are the same and that those of $\mathscr{S}_v(\mathbb{R}^n)$ and $\mathcal{S}_v(\mathbb{R}^n)$ are the same. It also follows from the results in §3.5.6 that the dimensions of $\mathcal{K}_v$ and $\mathcal{K}'_v$ are the same and that those of $\mathcal{S}_v(\mathbb{R}^n)$ and $\mathcal{S}'_v(\mathbb{R}^n)$ are the same.

Extensions to systems of linear forms are discussed in Chapters 4 and 5 and to hyperbolic space in Chapter 7. Restricted Diophantine approximation where the denominators are limited to certain sequences has been studied in [52], [54], [98], [192], see also [115, Chapter 10]. This allows monotonicity restrictions to be relaxed [75, 124].

Jarník's result [125] for $\mathfrak{B}$ was extended to systems $\mathfrak{B}(\mathbb{R}^{mn})$ of badly approximable linear forms by W. M. Schmidt who showed that $\mathfrak{B}(\mathbb{R}^{mn})$ is 'thick' in the sense that for each non-empty open subset W, $\dim \mathfrak{B}(\mathbb{R}^{mn}) \cap W = mn$ [200]. S. J. Patterson extended Jarník's result to a Fuchsian group setting [177, §10] (for more general results see §7.7). Generalisations to bounded orbits of flows on manifolds were developed by S. G. Dani [66, 67]. For related results in complex analysis see [104], [105]. Recently Kleinbock has extended Schmidt's result to systems of inhomogeneous linear forms [138].

§3.3 Hausdorff's normalising factor c_s is sometimes retained, see for example [103]. This makes no difference to the Hausdorff dimension and is an instance of the observation that 'constant factors do not matter' (of course they affect the measure).

Sets with positive and finite critical Hausdorff measure are called *s-sets* and have some nice properties, see [100]. The sets $\mathscr{K}_v$, $\mathfrak{B}$ and so on are not s-sets.

§3.4 Theorem 3.5 implies that the Hausdorff dimension of a non-measurable set in $\mathbb{R}^n$ is n.

When the approximation function $\psi(q)$ is much smaller than q^{-v}, the simple bound arising from the natural cover is not necessarily correct, see [81].

Diophantine approximation and Hausdorff dimension are discussed in [35];the little book [43] gives an account in Russian.

§3.5.4 The following contrapositive form is also used [48, 52, 54, 126]. Let s be any positive number with $s < h$. If any collection $\{C\}$ of arbitrarily small sets C with $\sum_C (\operatorname{diam} C)^s < \varepsilon$ for some $\varepsilon > 0$ cannot cover E, then $\dim E \geqslant h$.

Very often some lower bound can be obtained from a subset of known dimension. As an example the planar set $\mathscr{L}_v = \{x \in \mathbb{R}^2 \colon |\mathbf{q}\cdot x - p| < |\mathbf{q}|^{-v}$ infinitely often$\}$ contains the 'rational' lines $\{x \in \mathbb{R}^2 \colon \mathbf{q}\cdot x = p\}$ which have dimension 1, so that $\dim \mathscr{L}_v \geqslant 1$. But $\dim \mathscr{L}_v = 1+3/(v+1)$ for $v \geqslant 2$ [53] and indeed such simple observations will not usually give the correct lower bound; however see [53], [81].

§3.5.6 The set $A = \{\xi \in \mathbb{R} \colon w(\xi) > v\}$ has Hausdorff dimension $2/(v+1)$ when $v \geqslant 1$. Güting has pointed out in [113] that the sequence of subsets $A_r = \mathcal{K}'_{v+r^{-1}}$, $r = 1, 2, \ldots$, of A satisfies

$\dim A_r < \dim A$ but the complement of $(\bigcup_r A_r)$ in A is empty and so has dimension 0.

§**3.5.7** R. Kaufman [132] has obtained a stronger result than the Jarník-Besicovitch theorem by showing that $\mathscr{K}_v$ has a compact subset supporting a positive measure μ with Fourier-Stieltjes transform $\hat{\mu}$ satisfying

$$\hat{\mu}(u) = \int_{\mathscr{K}_v} e^{2\pi iux}\, d\mu(x) = o(\log(|u|))\, |u|^{v+1/2}$$

as $|u| \to \infty$.

CHAPTER 4

Upper bounds for Hausdorff dimension

4.1. Introduction

As was shown in the preceding chapter, the natural cover of a lim-sup set offers a general approach to the problem of obtaining upper bounds for the Hausdorff dimension of sets of ψ-approximable points. When the variables are independent, the determination of the correct upper bound for the Hausdorff dimension is often relatively simple. This is so for the Jarník-Besicovitch theorem (discussed in §3.5.2), for sets arising in the classification of transcendence of real and complex numbers [13, 142], for systems of linear forms and for approximation by algebraic numbers (treated in §4.1.1 and §5.2.1 below).

On the other hand, when the points under consideration lie on a manifold, the variables are dependent and the coordinates of the points are functionally related. This makes the calculation of the correct bounds much harder. Moreover when considering the set $\mathscr{L}(M\,;\psi)$ of dually ψ-approximable points on a manifold M, the measure of the elements associated with tangents in a natural cover can be very large. These problems are hardest for curves but have been resolved for planar curves and for the curve $\mathscr{V} = \{(t, t^2, \dots, t^n) : t \in \mathbb{R}\}$ (see §4.2.2 and §4.2.3 below for a brief discussion). Simultaneous Diophantine approximation is also very difficult and involves the distribution of rational points near the manifold. Exponential sum techniques however yield bounds.

When considering Hausdorff dimension, some of the simplifying assumptions made when considering Lebesgue measure have to be modified. Nevertheless the line of attack for upper estimates for the Hausdorff dimension remains much the same, with Lebesgue measure estimates being a starting point. We now give some illustrative examples, starting with systems of linear forms.

4.1.1. A general Jarník-Besicovitch theorem. The original theorem can be extended to systems of linear forms with a general approximation function ψ [82]. The correct upper bound for the Hausdorff dimension is obtained by decomposing the natural covering elements of thin parallelepipeds into hypercubes. Recall from §1.3 that $W(\psi)$ denotes the set of ψ-approximable points A (identified with $m \times n$ matrices) in $\mathbb{R}^{mn}$ which satisfy

$$|\langle \mathbf{q}\,A\rangle| < \psi(|\mathbf{q}|)$$

for infinitely many $\mathbf{q} \in \mathbb{Z}^m$. The *lower order* $\lambda(f)$ of a function $f\colon \mathbb{N} \to \mathbb{R}^+$ is defined by

$$\liminf_{N\to\infty} \frac{\log f(N)}{\log N}.$$

LEMMA 4.1. *Let $\psi\colon \mathbb{N} \to \mathbb{R}^+$ be a positive function and let the lower order of $1/\psi$ be λ. Then*

$$\dim W(\psi) \leqslant \begin{cases} (m-1)n + \frac{m+n}{\lambda+1}, & \lambda \geqslant m/n, \\ mn & \lambda \leqslant m/n. \end{cases}$$

PROOF. Theorem 3.5 implies that $\dim W(\psi) \leqslant mn$ since $W(\psi) \subset \mathbb{R}^{mn}$. For the other part of the bound, we note that the set $W(\psi)$ is invariant under translation by matrices in $\mathbb{Z}^{mn}$ and it follows that we can assume without loss of generality that $A \in [0,1]^{mn}$. Let $R_\mathbf{q}$ be the resonant set consisting of matrices $A \in [0,1]^{mn}$ such that $\langle \mathbf{q}\, A\rangle = 0$ and for each $\delta > 0$ let

$$B_\delta(R_\mathbf{q}) = \{A \in [0,1]^{mn} \colon |\langle \mathbf{q}\, A\rangle| < \delta\}.$$

Then

$$W(\psi) = \{A \in [0,1]^{mn} \colon A \in B_{\psi(|\mathbf{q}|)}(R_\mathbf{q}) \text{ for infinitely many } \mathbf{q} \in \mathbb{Z}^m\},$$

and the sets $B_{\psi(|\mathbf{q}|)}(R_\mathbf{q})$, $\mathbf{q} \in \mathbb{Z}^m$, are a natural cover for $W(\psi)$. To obtain a hypercube cover, we observe that each $B_{\psi(|\mathbf{q}|)}(R_\mathbf{q})$ can be covered by a collection $\mathscr{C}(\mathbf{q})$ say of

$$\ll \left(\frac{\psi(|\mathbf{q}|)}{|\mathbf{q}|}\right)^{-(m-1)n} |\mathbf{q}|^n \ll |\mathbf{q}|^{mn}\, \psi(|\mathbf{q}|)^{-(m-1)n}$$

mn-dimensional hypercubes C of sidelength $\ell(C) = 4\psi(|\mathbf{q}|)/|\mathbf{q}|$ and with centres on the $(m-1)n$-dimensional (resonant) set $R_\mathbf{q}$ consisting of a finite union of hyperplanes at integral multiples of $\psi(|\mathbf{q}|)/|\mathbf{q}|$ apart on the hyperplanes

$$R_{\mathbf{q},\mathbf{r}} = \{A \in [0,1]^{mn} \colon \mathbf{q}A = \mathbf{r}\}.$$

Moreover

$$\sum_{C\in\mathscr{C}} \ell(C)^s \leqslant 4^s \sum_{q=1}^{\infty} \sum_{|\mathbf{q}|=q} |\mathbf{q}|^{mn}\, \psi(|\mathbf{q}|)^{-(m-1)n} \left(\frac{\psi(|\mathbf{q}|)}{|\mathbf{q}|}\right)^s.$$

By the definition of lower order, $\psi(q) \ll q^{-\lambda+\varepsilon}$ for any $\varepsilon > 0$. Let

$$s = (m-1)n + (m+n)/(\lambda+1) + \eta$$

where $\eta > 0$ and let $\mathscr{C}$ be the union of the $\mathscr{C}(\mathbf{q})$ for $\mathbf{q} \neq 0$. Then $\mathscr{C}$ is a cover for $W(\psi)$ and

$$\sum_{C\in\mathscr{C}} \ell(C)^s \ll \sum_{q=1}^{\infty} q^{m-1+mn-s}\psi(q)^{s-(m-1)n} \ll \sum_{q=1}^{\infty} q^{m+n+1-(s-(m-1)n)(\lambda+1-\varepsilon)}$$
$$\ll \sum_{q=1}^{\infty} q^{-1-(\lambda+1)\eta+\varepsilon(m+n)/(\lambda+1)+\varepsilon\eta} < \infty$$

for ε sufficiently small. The result follows by the Hausdorff-Cantelli lemma. □

The complementary lower inequality is proved in Lemma 5.9.

4.1.2. Approximation by real algebraic numbers. This extension of the familiar rational approximation is an expanded version of [13, §2] with different notation. Let $\mathbb{A}_n$ denote the set of real algebraic numbers α of degree at most $n \geqslant 2$. For each real number ξ and positive integer n, let $\omega(\xi) = \omega^{(n)}(\xi)$ be the supremum of positive numbers w for which the inequality

$$|\xi - \alpha| < h(\alpha)^{-w}$$

has infinitely many solutions in algebraic numbers $\alpha \in \mathbb{A}_n$ ($h(\alpha)$ is the height of the minimal polynomial for α). Let

$$\mathfrak{K}_w = \mathfrak{K}_w^{(n)} = \{\xi \in \mathbb{R} : \omega(\xi) \geqslant w\} \tag{4.1}$$

(in [13], the set $\mathfrak{K}_w$ is written $K_n(\lambda)$, where $\lambda = w/(n+1)$). It is proved in [13] that $\dim \mathfrak{K}_w = \dim \mathfrak{K}'_w = (n+1)/w$ for $w > n+1$, where $\mathfrak{K}'_w$ is the set of ξ for which $\omega(\xi) = w$. As a further illustration [22], the correct upper bound is now obtained (the complementary lower bound is obtained in §5.2.1).

For each $j = 1, \ldots, n$ and each positive integer k, denote by $\mathbb{A}_n(j,k)$ the set of real algebraic numbers α in $\mathbb{A}_n$ with height k and with the coefficient a_j of x^j in the minimal polynomial for α having modulus k. The cardinality of the set $\mathbb{A}_n(j,k)$ is at most $2(2k+1)^n$ and $\mathbb{A}_n$ can be represented as

$$\mathbb{A}_n = \bigcup_{j=0}^{n} \bigcup_{k=1}^{\infty} \mathbb{A}_n(j,k).$$

For fixed real $\alpha \in \mathbb{A}_n(j,k)$ and arbitrary $\varepsilon > 0$ the inequality

$$|\xi - \alpha| < k^{-w+\varepsilon}$$

holds for each $\xi \in I(\alpha, k^{-w+\varepsilon})$, where $I(\alpha,\delta) = (\alpha - \delta, \alpha + \delta)$. Given $\varepsilon > 0$,

$$\mathfrak{K}_w \subseteq \{\xi \in \mathbb{R} : \xi \in I(\alpha, h(\alpha)^{-w+\varepsilon}) \text{ for infinitely many } \alpha \in \mathbb{A}_n\}.$$

The s-length of this cover is at most

$$2^s \sum_{\alpha\in\mathbb{A}_n} h(\alpha)^{-s(w-\varepsilon)} \ll \sum_{k=1}^{\infty}\sum_{j=0}^{n} \sum_{\alpha\in A_n(j,k)} k^{-(w-\varepsilon)s} \ll \sum_{k=1}^{\infty} k^{-(w-\varepsilon)s+n} < \infty$$

for $s > (n+1)/(w-\varepsilon)$, whence by the Hausdorff-Cantelli lemma (Lemma 3.10),

$$\dim \mathfrak{K}_w \leqslant (n+1)/(w-\varepsilon)$$

and since $\varepsilon > 0$ is arbitrary, $\dim \mathfrak{K}_w \leqslant (n+1)/w$. Note that this estimate is non-trivial only for $w > n+1$.

Next we consider simultaneous approximation involving different approximation functions of points in the plane by real algebraic points [43, Chapter 2, §4]. These functions do not appear explicitly in Lebesgue measure questions but they can affect the Hausdorff dimension. To give an extreme example, if $v > 1/2$, each of the two inequalities

$$\|qx\| \cdot \|qy\| < q^{-2v} \quad \text{and} \quad \max\{\|qx\|, \|qy\|\} < q^{-v}$$

has infinitely many solutions in positive integers q for almost no $(x, y) \in \mathbb{R}^2$ ([108], [205] and by Khintchine's theorem). But the Hausdorff dimension of the set of points (x, y) satisfying the first inequality infinitely often is $1 + 2/(2v+1)$ for $v \geqslant 1/2$ [53, 81], while that of the set $\mathscr{S}_v(\mathbb{R}^2)$ of points satisfying the second infinitely often for $v \geqslant 1/2$ is $3/(1+v)$ by Jarník's theorem [127].

Let $n \geqslant 2$ and let P be an integral irreducible polynomial of degree at most n. The Hausdorff dimension of the set $\mathfrak{K}(w_1, w_2)$ of vectors (ξ_1, ξ_2) in the plane for which the inequalities

$$|\xi_1 - \alpha_1| < h(P)^{-w_1}, \quad |\xi_2 - \alpha_2| < h(P)^{-w_2}, \tag{4.2}$$

where α_1, α_2 are two real zeros of P, hold for infinitely many P will be determined. As usual the upper and lower bounds are obtained separately.

LEMMA 4.2. *Assume that* $w_1 \geqslant w_2 \geqslant 1$ *and* $w_1 + w_2 > n+1$. *Then*

$$\dim \mathfrak{K}(w_1, w_2) \leqslant \begin{cases} (n+1+w_1-w_2)/w_1, & 1 \leqslant w_2 < n+1, \\ (n+1)/w_2, & w_2 \geqslant n+1. \end{cases}$$

Note that for $w_2 \geqslant n+1$, the estimate is independent of w_1.

PROOF. There is no loss of generality in considering points (ξ_1, ξ_2) in $[0,1]^2$. By hypothesis, $w_1 > (n+1)/2$. Fix the height of the polynomials P to be N where $N \geqslant N_0$, a sufficiently large integer. The set of points (ξ_1, ξ_2) satisfying the inequalities (4.2) above when α_1, α_2 are two real roots of the polynomial P consists of $\ll N^n$ rectangles $R(N, (\alpha_1, \alpha_2))$ say, centred at (α_1, α_2) and of area $4N^{-w_1-w_2}$.

The rectangle $R(N, (\alpha_1, \alpha_2))$ can be covered by $\ll N^{w_1-w_2}$ squares S_j with sidelength $2N^{-w_1}$. The collection of all the S_j which cover all the rectangles $R(N, (\alpha_1, \alpha_2))$ with $N \geqslant N_0$ is a natural cover of $\mathfrak{K}(w_1, w_2)$ by squares with s-length

$$\sum_{S_j} \ell(S_j)^s \ll \sum_{N \geqslant N_0} N^{n+w_1-w_2} (2^{3/2} N^{-w_1})^{(n+1+w_1-w_2+\delta)/w_1} \ll \sum_{N \geqslant N_0} N^{-1-\delta} < \infty$$

when $s = (n+1+w_1-w_2+\delta)/w_1$ for any $\delta > 0$. Then by the Hausdorff-Cantelli lemma and since δ is arbitrary,

$$\dim \mathfrak{K}(w_1, w_2) \leqslant (n+1+w_1-w_2)/w_1.$$

If $w_2 \geqslant n+1$ then $\dim \mathfrak{K}(w_1, w_2) \leqslant 1$. However, using the larger squares with sidelength N^{-w_2} we obtain, as above,

$$\dim \mathfrak{K}(w_1, w_2) \leqslant (n+1)/w_2.$$

The other case $w_1 \leqslant w_2$ follows by an obvious change in the statement. □

The downward inequality is done in §5.2.1.

4.1.3. Complex functions, capacity and Hausdorff dimension. A natural extension of the above is the approximation of numbers in $\mathbb{C}$ by complex algebraic numbers. For each complex number z, let $\omega^*(z)$ be the supremum of those positive numbers w for which the inequality

$$|z-\alpha| < h(\alpha)^{-w}$$

has infinitely many solutions in algebraic numbers α, $\deg \alpha \leqslant n$. Let

$$A_w = \{z \in \mathbb{C} : \omega^*(z) \geqslant w\}.$$

As above, there is no loss in generality in restricting z to the unit square in $\mathbb{C}$. This set can be covered by discs centred at algebraic numbers in the square and with suitable radii. It can be shown on the lines of the above example that $\dim A_w \leqslant (n+1)/w$. This estimate is non-trivial when $w > (n+1)/2$; the estimates in this and the other examples for ω and ω^* are natural in the light of Wirsing's conjecture [232] that $\omega(\xi) \geqslant n+1$ when $\xi \in \mathbb{R}$.

Sets of the form $B(f;\varepsilon) = \{z \in \mathbb{C} : |f(z)| < \varepsilon\}$, where ε is positive and f is a complex function, occur often. Information about the capacity of $B(f;\varepsilon)$ can be used to discuss the Hausdorff dimension, as is done for example in [117] and [184]. One result which we use, namely that the (Euclidean) diameter of a continuum (a compact connected Hausdorff space) E is at most $4\,\mathrm{C}_0(E)$, follows readily from the definition of the capacity $\mathrm{C}_0(E)$ of E (see §3.5.7). Now consider the polynomial $P(z) = a_n z^n + a_{n-1}z^{n-1} + \cdots + a_0$ of degree n. By modifying Problem 5 in [184, Chapter 11] and using the homogeneity of the capacity, it can be seen that

$$\mathrm{C}_0(B(P;\varepsilon)) = \left(\frac{\varepsilon}{|a_n|}\right)^{1/n}.$$

This can be used to obtain an upper bound of the Hausdorff dimension of the set

$$\mathfrak{M}_v^* = \{z \in \mathbb{C} \colon |P(z)| < h(P)^{-v} \text{ for infinitely many integral } P\}$$

($\mathfrak{M}_v^*$ is the complex version of $\mathfrak{M}_v$). For a fixed polynomial P, the set of points z satisfying $|P(z)| < h(P)^{-v}$ consists of at most n bounded connected domains by continuity and when considering the Hausdorff dimension, it suffices to consider any one of them. By arguing as in [13] or [208, Chapter 1], it also suffices to consider the subset $\mathcal{M}_v^*$ of $\mathfrak{M}_v^*$ in which the integral polynomials P are leading, irreducible and of degree n, *i.e.*, $P \in \mathfrak{I}_n$ (see §2.4.1). Now $\mathcal{M}_v^*$ has a natural cover consisting of the sets $B(P\,; h(P)^{-v})$, $P \in \mathfrak{I}_n(N)$ (*i.e.*, $h(P) = N$), $N = 1, 2, \ldots$, with diameter $O\left(N^{-(v+1)/n}\right)$. The s-length of this cover satisfies

$$\sum_{P\in\mathfrak{I}_n} (\operatorname{diam} B(P\,; h(P)^{-v}))^s \ll \sum_{N=1}^{\infty} \sum_{P\in\mathfrak{I}_n(N)} (h(P)^v N)^{-s/n}$$
$$\ll \sum_{N=1}^{\infty} N^{-(1+v)s/n}\, N^n < \infty$$

for $s > n(n+1)/(v+1)$. Hence the Hausdorff-Cantelli lemma implies that $\dim \mathfrak{M}_v^* \leqslant n(n+1)/(v+1)$ when $v \geqslant n(n+1)-1$. Thus estimates for the capacity of the set can give weak but non-trivial information about Hausdorff dimension. Sharper bounds can be obtained however in special cases (see §4.2.1 below) or with the help of estimates from the theory of transcendental numbers [13, 33].

4.2. Diophantine approximation on manifolds

We now turn to the more general question of Diophantine approximation on manifolds. Recall from §3.6 that to determine the Hausdorff dimension of a set on a manifold M, it suffices to work in a parametrisation domain U which is a suitable small hypercube.

4.2.1. The parabola. We begin by considering the simplest case of dependent variables, namely the set $\mathscr{L}_v(\mathscr{V}^{(2)})$ of points on the parabola which are dually v-approximable. In §4.1.1 and §4.1.2, the diameters of elements in the covers are 'small'. In this example some of the elements in the cover have a 'large' diameter but sufficiently rarely to allow the Hausdorff-Cantelli lemma to give the correct upper bound for the Hausdorff dimension. The analysis is an adaptation of §2.4.4 where the quadratics to be considered were taken to be irreducible and leading and the variable $t \in (-3, 3)$.

The reduction from general integer polynomials to leading irreducible ones when considering the Hausdorff dimension of the set $\mathfrak{M}_v$ is very similar to that when considering Lebesgue measure; a brief discussion is in §4.2.3 below. Accordingly we will show that the set $\mathcal{Q}_v = \mathcal{M}_v^{(2)}$ of $t \in (-3, 3)$ such that

$$|Q(t)| < h(Q)^{-v} \tag{4.3}$$

for infinitely many Q in $\mathfrak{I}_2$, the set of irreducible leading integral quadratics $Q(t) = h(Q)t^2 + a_1 t + a_0$, has Hausdorff dimension at most $3/(v+1)$ when $v \geqslant 2$. This implies that $\dim \mathfrak{Q}_v \leqslant 3/(v+1)$ for $v \geqslant 2$ (this estimate is non-trivial only when $v > 2$). In view of §1.5.3, the inequality (4.3) is closely related to the inequality

$$\|a_1 t + a_2 t^2\| < |(a_1, a_2)|^{-v}$$

and indeed the sets $\mathscr{L}_v(\mathscr{V}^{(2)})$, $\mathcal{Q}_v$ and $\mathfrak{Q}_v$ have the same Hausdorff dimension.

We start by expressing $\mathcal{Q}_v$ in the usual lim-sup form:

$$\mathcal{Q}_v = \{t \in [0,1] \colon t \in J_Q \text{ for infinitely many } Q \in \mathfrak{I}_2\},$$

where each element J_Q in the natural cover is the union of two intervals I_Q, I'_Q each of length at most $2N^{-v}(\Delta(Q))^{-1/2}$ and which coincide (and may be empty) if Q has non-real roots (see §7.5). Then for $s \in [0,1]$, the s-length of the natural cover for $\mathcal{Q}_v$ by the intervals I_Q, I'_Q is

$$\ll \sum_{N=1}^{\infty} \sum_{Q} N^{-vs} \Delta(Q)^{-s/2} \ll \sum_{N=1}^{\infty} N^{-vs} \sum_{|a_1| \leqslant N} \sum_{a_0} |a_1^2 - 4a_0 N|^{-s/2},$$

where the second sum on the left hand side is over quadratics $Q \in \mathfrak{I}_2$ of height N and the last sum on the right hand side is over integers $a_0 \neq a_1^2/4N$ with $|a_0| \leqslant N$.

But for each fixed a_1,

$$\sum_{a_0=-N}^{N} |a_1^2 - 4a_0 N|^{-s/2} \ll 1 + \sum_{k=1}^{N} (kN)^{-s/2} \ll N^{-s/2} N^{1-s/2} \ll N^{1-s},$$

whence

$$\sum_{N=1}^{\infty} \sum_{Q} |J_Q|^s \ll \sum_{N=1}^{\infty} N^{-vs+2-s} < \infty,$$

provided $s > 3/(v+1)$. It now follows from the Hausdorff-Cantelli lemma that $\dim \mathcal{Q}_v \leqslant 3/(v+1)$ for $v > 2$ whence $\dim \mathfrak{Q}_v \leqslant 3/(v+1)$.

The general downward inequality for $\dim \mathfrak{M}_v^{(n)}$, $n \geqslant 1$, was established by Baker and Schmidt using the idea of regular systems, introduced in [13] (see §5.2).

4.2.2. Planar curves. We consider briefly the dual form of Diophantine approximation on a C^3 planar curve Γ given by

$$\{x(s) \in \mathbb{R}^2 \colon s \in I\},$$

where I is an interval and s is arclength (and is of course unrelated to the exponent s elsewhere). In [198] W. M. Schmidt proved that when the curvature of Γ is non-zero almost everywhere on I, Γ is extremal, *i.e.*, the set $\mathscr{L}_v(\Gamma)$ of points $x(s)$ such that $\|\mathbf{q} \cdot x(s)\| < |\mathbf{q}|^{-v}$ holds for infinitely many $\mathbf{q} \in \mathbb{Z}^2$ is null for $v > 2$ (see §2.2). R. C. Baker [16] refined and extended Schmidt's result. Let $z_1, z_2 \in \mathbb{C}$ and let $\omega_{\mathscr{L}}(z_1, z_2)$ be the supremum of those $w \in \mathbb{R}$ for which the inequality

$$|q_1 z_1 + q_2 z_2 - p| < |(q_1, q_2, p)|^{-w} \tag{4.4}$$

holds for infinitely many $q_1, q_2, p \in \mathbb{Z}$. The set $\{s \in I \colon \omega_{\mathscr{L}}(x(s)) \geqslant v\} = X(\Gamma, v)$ in the notation of [16] corresponds to taking $(z_1, z_2) = (x_1(s), x_2(s)) = x(s) \in \mathbb{R}^2$. In proving equality in [16], he showed that if the curvature of Γ was non-zero except for a set of Hausdorff dimension 0, then

$$\dim X(\Gamma, v) \leqslant 3/(v+1) \tag{4.5}$$

for $v \geqslant 2$ (this extends his earlier results for the parabola [15]). It follows that the Hausdorff dimension of $\mathscr{L}_v(\Gamma)$ and $\mathcal{L}_v(\Gamma)$ is at most $3/(v+1)$ for $v \geqslant 2$. As has been pointed out in §3.6, it suffices to consider the set (related to $X(\Gamma, v)$)

$$L_v = \{s \in I \colon \|\mathbf{q} \cdot x(s)\| < |\mathbf{q}|^{-v} \text{ for infinitely many } \mathbf{q} \in \mathbb{Z}^2\}$$

of points s on a small interval I on which it may be assumed without loss of generality that the (continuous) curvature $\kappa(s)$ and the components $x_1'(s)$, $x_2'(s)$ of the derivative $x'(s)$ are bounded away from 0. The proof is long and involved, resting on W. M. Schmidt's results in [198] and [199]. An estimate for the number of solutions of $\|g(x,y)\| \ll x^{-\nu}$ for a certain homogeneous function g in [199] is an essential result and is discussed in [210, Chapter 1, Lemma 2].

4.2.3. Curves in higher dimensions. Although Diophantine approximation on curves is difficult, correct upper bounds for the Hausdorff dimension of sets of ψ-approximable points on curves in $\mathbb{R}^n$ for $n > 2$ are known in a few cases, including $\mathscr{V}$. The analysis of the Hausdorff dimension of $\mathscr{M}_v$ can be simplified along the lines of §2.4 in [208, Chapter 1] or [33]. It suffices to consider points t for which the inequality

$$|P(t)| < h(P)^{-v}$$

holds for infinitely many irreducible integral polynomials P of degree n. Next, each such polynomial P can be transformed to a leading polynomial by a translation and an inversion without affecting the Hausdorff dimension (see Lemmas 2, 3, 4 in [15]). Thus only leading irreducible polynomials and the set $\mathcal{M}_v$ need be considered (further discussion is in the references cited). Recall that $\mathscr{L}_v(\mathscr{V})$, $\mathfrak{M}_v$ and $\mathscr{M}_v$ (see §1.5.3) are closely related.

In [13] Baker and Schmidt established the inequalities

$$(n+1)/(v+1) \leqslant \dim \mathscr{M}_v < 2(n+1)/(v+1)$$

when $v \geqslant n$ for each integer $n \geqslant 1$. They used regular systems (see §5.2) and the inclusion $\mathscr{M}_v \subset \mathfrak{K}_{v+1}$ to obtain the lower bound and conjectured that when $n \geqslant 1$ and $v \geqslant n$,

$$\dim \mathscr{M}_v = \frac{n+1}{v+1}. \tag{4.6}$$

The case $n = 1$ corresponds to the Jarník-Besicovitch theorem. F. Kasch and B. Volkmann [131] obtained the upper bound $3/(v+1)$ for $n = 2$ which, when combined with the lower bound in [131], implied the Baker-Schmidt conjecture in this case. Later R. C. Baker [15] used A. Baker's refinement [10] of Sprindžuk's theorem and ideas of Volkmann [228] to establish equality for $n = 3$ in the real case and also for $n = 4$ and 5 in the non-real case. He also obtained the exact dimension for large v (about n^2). The real case (4.6) was finally proved for any degree by Bernik [33, 43] and implies that the Hausdorff dimension of the closely related sets $\mathfrak{M}_v$, $\mathfrak{M}_v'$, $\mathscr{L}_v(\mathscr{V})$ and $\mathscr{L}_v'(\mathscr{V})$ (see 2.4) is also $(n+1)/(v+1)$ for $v \geqslant n$. Although it follows similar lines to that of Sprindžuk's proof of Mahler's conjecture [208] discussed in §2.2, Bernik's proof is considerably more complicated and will be omitted. One of the basic ideas, also used in Chapter 2 (Lemma 2.9), is that two relatively prime integral polynomials can only have small absolute values (in terms of their heights and degrees) simultaneously on a very short interval.

4.3. Smooth manifolds of dimension at least 2

We now consider the Hausdorff dimension of the set of ψ-approximable points on manifolds of dimension at least 2 and subject to mild geometric and analytic constraints. In view of Lemma 3.6 and the invariance of the Hausdorff dimension of a set under a bi-Lipschitz map (Theorem 3.7), we can work, as we did in Chapter 2 for Lebesgue measure, with $L(\psi)$ or $S(\psi)$ in the parametrisation domain U for the manifold M instead of with $\mathscr{L}(M;\psi)$ or $\mathscr{S}(M;\psi)$. To discuss the Hausdorff dimension of $L(\psi)$, the covers used in the Lebesgue measure estimates obtained in Chapter 2 could be subdivided into small hypercubes. However, a slightly weaker curvature condition is imposed and the correct upper bound for the Hausdorff dimension obtained. This is done in [89] but only for the case $n=1, m=2$. Simultaneous Diophantine approximation is less straightforward but some estimates can be obtained using exponential sum techniques.

4.3.1. Another curvature condition.

We again assume that M is a C^3 m-dimensional submanifold of $\mathbb{R}^n$. The manifold M will be said to be 2-*curved* at $x \in M$ if for any unit vector $\nu \in T_xM^{\perp}$, at least two of the principal curvatures $\kappa_i(x,\nu)$ do not vanish. (The condition of 2-convexity at x introduced in §2.5.1 requires in addition that the two principal curvatures have the same sign.) This curvature condition allows the correct upper bound for the Hausdorff dimension of $\mathscr{L}_v(M)$ to be obtained. For a surface in $\mathbb{R}^3$, it reduces to the Gaussian curvature being non-zero but it is weaker for $m \geqslant 3$. Recall that it suffices to take the Monge parametrisation domain U to be a suitable small hypercube in $[-1,1]^m$.

The set M_0 consisting of points $x \in M$ at which all or all but one of the principal curvatures vanish for some unit vector $\nu \in T_\xi M^{\perp}$ has a simple algebraic description in terms of the auxiliary function $\Phi_\omega\colon \mathbb{R}^k \to \mathbb{R}$ given by $\Phi_\omega(u) = \omega \cdot \varphi(u)$ (see §2.5.1). Let U_0 be the set of $u \in U$ for which the rank of the Hessian matrix $\operatorname{Hess}\Phi_\omega(u)$ is at most 1 for some vector $\omega \in \mathbb{R}^k$. Then

$$M_0 = \operatorname{graph}(\varphi\colon U_0 \to \mathbb{R}^k)$$

since there is at most one non-zero principal curvature at $x=(u,\varphi(u))$ for some unit vector $\nu \in T_xM^{\perp}$ if and only if the matrix $\operatorname{Hess}\Phi_\omega(u)$ has at most one non-zero eigenvalue for some non-zero $\omega \in \mathbb{R}^k$ (see §2.5.1).

Determining the upper bound involves rather elaborate arguments but the idea is based on the construction of a suitable cover and the application of the Hausdorff-Cantelli lemma. First some preliminary estimates are needed. These are based on [16, §3] and [199].

LEMMA 4.3. *Let $f: I \to \mathbb{R}$ be a C^2 function defined on the interval I and suppose that for each v in an interval $J \subset \mathbb{R}$ there is just one point $u = u(v)$ such that $f(u) = v$ and $f'(u(v)) \neq 0$. Then the set $f^{-1}(J)$ is a subinterval of I of length*

$$|f^{-1}(J)| = \int_J |f'(u(v))|^{-1}\, dv.$$

PROOF. From the hypotheses, the inverse function $f^{-1}: J \to I$ exists and is differentiable. Now

$$\begin{aligned}|f^{-1}(J)| &= \int_I \chi_{f^{-1}(J)}(u)du = \int_{f(I)} \chi_{f^{-1}(J)}(f^{-1}(v))|f'(u(v))|^{-1}\, dv \\ &= \int_J |f'(u(v))|^{-1}\, dv\end{aligned}$$

on changing the variable u to $f^{-1}(v)$. □

This is Hilfsatze 3 of [199] and it immediately implies the following lemma.

LEMMA 4.4. *Let $f: I \to \mathbb{R}$ be a C^2 function on the interval I and suppose that $|f'(u)| \geqslant \delta > 0$ for each $u \in I$. Then for any $\rho > 0$, the set $f^{-1}(-\rho, \rho)$ is an interval (possibly empty) and*

$$|f^{-1}(-\rho, \rho)| \leqslant 2\rho/\delta.$$

The next estimate is also drawn from [199] and gives the measure of the inverse image of a neighbourhood of the origin when the second derivative is non-zero.

LEMMA 4.5. *Let $f: \mathbb{R} \to \mathbb{R}$ be C^2 and suppose that there exists a constant $c > 0$ such that $|f''(u)| \geqslant c$ for each $u \in \mathbb{R}$. Then f' is either strictly increasing or strictly decreasing and so vanishes just once, at $u = u_0$ say. For any $\rho > 0$, the set $f^{-1}(-\rho, \rho)$ is a union of at most two intervals, with*

$$|f^{-1}(-\rho, \rho)| \leqslant 8c^{-1/2} \min\{|f(u_0)|^{-1/2}\rho, \rho^{1/2}\}.$$

PROOF. Write $f_0 = f(u_0)$ and assume as we may that $f''(u) > 0$ for each $u \in \mathbb{R}$ (the other case is entirely similar). Let

$$\begin{aligned}I_+ &= (f^{-1}(-\rho, \rho)) \cap [u_0, \infty), \\ I_- &= (f^{-1}(-\rho, \rho)) \cap (-\infty, u_0).\end{aligned}$$

Since f' is positive on (u_0, ∞), the function $v = f(u)$ has a C^2 inverse $u = f^{-1}(v)$ on (f_0, ∞). Hence I_+ is an interval and similarly I_- is also an interval.

Now $|f'(u)|^2 \geqslant 2c\,|f(u) - f_0| = 2c\,|v - f_0|$ for $u > u_0$ since on differentiating $(f'(u(v)))^2$ with respect to $v \in (f_0, \infty)$, we get

$$\frac{d}{dv} f'(u(v))^2 = 2f'(u(v))\, f''(u(v))\, \frac{du(v)}{dv} = 2f''(u(v)),$$

whence $|d(f'(u(v))^2)/dv| \geqslant 2c$ for each $v > f_0$. Next suppose that $|f_0| \geqslant 2\rho$. If $|f(u)| < \rho$, then

$$|f(u) - f_0| \geqslant |f_0| - |f(u)| > |f_0|/2$$

and $|f'(u)| > (c\,|f_0|)^{1/2}$ for each $u \in I_+$. Hence by Lemma 4.4, $|I_+| \leqslant 2\rho(c\,|f_0|)^{-1/2}$ and the same estimate holds for I_-. Thus

$$|f^{-1}(-\rho,\,\rho)| \leqslant 4\rho(c\,|f_0|)^{-1/2}.$$

When $|f_0| < 2\rho$, by Lemma 4.3,

$$|I_+| \leqslant \int_{-\rho}^{\rho} (2c\,|v - f_0|)^{-1/2}\, dv \leqslant 2^{3/2}(\sqrt{3}-1)(\rho/c)^{1/2}.$$

The same estimate holds for I_- and the lemma follows. □

Now let $h\colon \mathbb{R}^2 \to \mathbb{R}$ be a C^3 function which for some $c > 0$ satisfies

$$\left|\frac{\partial^2 h(u,v)}{\partial u^2}\right| \geqslant c, \quad \left|\frac{\partial^2 h(u,v)}{\partial v^2}\right| \geqslant c, \quad \left|\frac{\partial^2 h(u,v)}{\partial u \partial v}\right| \leqslant c/2, \tag{4.7}$$

for each $(u,v) \in \mathbb{R}^2$. These conditions ensure that when ε is small enough, the diagonal terms in the quadratic term in the Taylor expansion of h dominate the off-diagonal terms. Given $v \in \mathbb{R}$, the function $\partial h(\cdot, v)/\partial u$ has derivative of positive modulus and so there exists a unique point $u_0(v)$ say for which $\partial h(u_0(v), v)/\partial u$ vanishes. By the implicit function theorem, u_0 is a C^2 function and so the function $h_0\colon \mathbb{R} \to \mathbb{R}$ defined by

$$h_0(v) = h(u_0(v), v) \tag{4.8}$$

is also C^2. Moreover $|h_0''(v)| \geqslant c/2$ for each $v \in \mathbb{R}$; it further follows that there is a unique point v_0 such that $h_0'(v_0) = 0$. We will write

$$h_{00} = h_0(v_0) = h(u_0(v_0), v_0). \tag{4.9}$$

4.3.2. A hypercube cover for $L(\psi)$. We consider the local Monge parametrisation (U, θ), where $\theta = 1_U \times \varphi$, for the manifold M (see §1.5.1). Recall that the set $L(\psi)$ is the local projection of the ψ-approximable points on M_U, *i.e.*,

$$\begin{aligned} L(\psi) &= \{u \in U\colon \|\mathbf{q}\cdot\theta(u)\| < \psi(|\mathbf{q}|) \text{ for infinitely many } \mathbf{q} \in \mathbb{Z}^n\} \\ &= \{u \in U\colon u \in B_{\psi(|\mathbf{q}|)}(\mathbf{q}) \text{ for infinitely many } \mathbf{q} \in \mathbb{Z}^n\} \end{aligned}$$

where $B_\delta(\mathbf{q}) = \{u \in U\colon |\langle \mathbf{q}\cdot\theta(u)\rangle| < \delta\}$. The family of sets $\{B_{\psi(|\mathbf{q}|)}(\mathbf{q})\}$, where $\mathbf{q} \in \mathbb{Z}^n \setminus \{0\}$, is a natural cover for $L(\psi)$ and is decomposed into a cover consisting of hypercubes in $\mathbb{R}^n$. Each element $B_{\psi(|\mathbf{q}|)}(\mathbf{q})$ in the natural cover can be expressed as

$$B_{\psi(|\mathbf{q}|)}(\mathbf{q}) = \bigcup_{p \in \mathbb{Z}} B_{\psi(|\mathbf{q}|)}(p, \mathbf{q})$$

where $B_\delta(p, \mathbf{q}) = \{u \in U\colon |\mathbf{q}\cdot\theta(u) - p| < \delta\}$ (given $\mathbf{q}$ and $\delta \leqslant 1/2$, there are $O(|\mathbf{q}|)$ integers p for which $B_\delta(p, \mathbf{q})$ is not empty).

The s-length of this cover is estimated on lines similar to those in §2.5.1. In particular the auxiliary function $g\colon U \to \mathbb{R}$ (2.44) given by

$$g(u) = \mathbf{q}\cdot\theta(u) - p = \mathbf{q}^{(1)}\cdot u + \mathbf{q}^{(2)}\cdot\varphi(u) - p$$

and its smooth extension to $\mathbb{R}^m$ are used. Recall from (1.21) that

$$K_{ir} = \sup\{|\partial\varphi_r/\partial u_i| : u \in U\}$$

and that $K = \max\{K_{ir} : 1 \leqslant i \leqslant m, 1 \leqslant r \leqslant k\}$.

LEMMA 4.6. *For each non-zero* $\mathbf{q} \in \mathbb{Z}^n$, $B_{\psi(|\mathbf{q}|)}(\mathbf{q})$ *has a hypercube cover* $\mathscr{C}(\mathbf{q})$ *such that* $\ell(C) = \psi(|\mathbf{q}|)/|\mathbf{q}|$ *for each* C *in* $\mathscr{C}(\mathbf{q})$ *and*

$$\ell^s(\mathscr{C}(\mathbf{q})) \ll \psi(|\mathbf{q}|)^{s-m+1}|\mathbf{q}|^{-s+m}|\log(\psi(|\mathbf{q}|)/|\mathbf{q}|)|.$$

PROOF. We begin by constructing a hypercube cover for $B(p, \mathbf{q}\,; \delta)$. This is done in two parts. First consider the 'transverse' case $|\mathbf{q}^{(2)}| \leqslant |\mathbf{q}|/(2kK)$. In this case, $|\mathbf{q}^{(1)}| = |\mathbf{q}|$ and $|q_i^{(1)}| = |\mathbf{q}|$ for $i = 1$ say by relabelling. By (2.46),

$$\left|\frac{\partial g(u)}{\partial u_1}\right| \geqslant |q_1| - \left|\mathbf{q}^{(2)} \cdot \frac{\partial \varphi(u)}{\partial u_1}\right| \geqslant \frac{1}{2}|\mathbf{q}|$$

for each $u \in U$. Let

$$\psi(|\mathbf{q}|)/|\mathbf{q}| = \eta(|\mathbf{q}|) \tag{4.10}$$

and for each $u \in U$, write $u = (u_1, u')$ where $u' = (u_2, \dots, u_m) \in \mathbb{R}^{m-1}$. For each $u' \in \mathbb{R}^{m-1}$ let

$$E(u') = \{u_1 \in \mathbb{R} : (u_1, u') \in U \text{ and } |g(u_1, u')| < 2\,\psi(|\mathbf{q}|)\}.$$

Given fixed u', $g(u_1, u')$ is a C^3 real valued function of the real variable u_1. Hence by the above bound on $|\partial g(u)/\partial u_1|$ and by Lemma 4.4, $E(u')$ is an interval of length $|E(u')| \leqslant 8\eta(|\mathbf{q}|)$. Hence $B_{\psi(|\mathbf{q}|)}(p, \mathbf{q}) = g^{-1}(-\psi(|\mathbf{q}|), \psi(|\mathbf{q}|)) \cap U$ is a neighbourhood of $R_{p,\mathbf{q}} = g^{-1}(0)$ and by the choice of U, $R_{p,\mathbf{q}}$ is close to being the intersection of an $(m-1)$-dimensional hyperplane with U. Hence $B_{\psi(|\mathbf{q}|)}(p, \mathbf{q})$ is close to a parallelepiped $I_1 \times I^{m-2}$ where I_1 is an interval of length $2\eta(|\mathbf{q}|)$ and I an interval of length ε. Consequently $B_{\psi(|\mathbf{q}|)}(p, \mathbf{q})$ can be covered by $\ll \eta(|\mathbf{q}|)^{-m+1}$ hypercubes C of sidelength $\ell(C) = \eta(|\mathbf{q}|)$.

A more precise argument goes as follows. Let

$$F(u') = \{w \in U : \operatorname{dist}(w, (E(u'), u')) < \eta(|\mathbf{q}|)\},$$

where $(E(u'), u') = \{(u_1, u') : u_1 \in E(u')\}$, be a section of $B_{\psi(|\mathbf{q}|)}(p, \mathbf{q})$, a neighbourhood of the resonant set $R_{p,\mathbf{q}} = \{u \in U : \mathbf{q} \cdot \varphi(u) = p\}$. Then the set $F(u')$ is a neighbourhood of $(E(u'), u')$. Thus for each u', the set $F(u')$ can be covered by a collection $\mathscr{C}(u')$ of at most $\ll 1$ hypercubes C with sidelength $\ell(C) = \eta(|\mathbf{q}|)$.

Next we construct closely spaced lattices centred at the origin. Let

$$\Lambda^1 = \left\{-1 + \frac{r}{[\rho^{-1}]} : r = 0, \dots, [\rho^{-1}]\right\} \cap [-1, 1]$$

and let Λ^i be the i-fold Cartesian product of Λ^1, so that Λ^i consists of $\ll \rho^{-i}$ points in $U \subset [-1,1]^i$. Let the spacing $\rho = \min\{\eta(|\mathbf{q}|)/(4m^{1/2}kK), \varepsilon/k^{1/2}\}$. By the definition of $\mathscr{C}(u')$, the collection

$$\mathscr{C}(p,\mathbf{q}) \ll \bigcup_{u' \in \Lambda^{m-1}} \mathscr{C}(u')$$

of $\ll \eta(|\mathbf{q}|)^{-m+1}$ hypercubes C has s-length

$$\ell^s(\mathscr{C}(p,\mathbf{q})) \ll \sum_{C \in \mathscr{C}(p,\mathbf{q})} \ell(C)^s \ll \eta(|\mathbf{q}|)^{s-m+1}$$

and covers the set $\bigcup_{u' \in \Lambda^{m-1}} F(u')$. Thus the collection covers $B_{\psi(|\mathbf{q}|)}(p,\mathbf{q})$, since given any point $w \in B_{\psi(|\mathbf{q}|)}(p,\mathbf{q})$ there exists a point $u \in U$ with

$$|w-u|_2 \leqslant \psi(|\mathbf{q}|)(2m^{1/2}|\mathbf{q}|kK)^{-1}$$

and $u' \in \Lambda^{m-1}$. Using the mean value theorem, we get that

$$|g(u)| \leqslant |g(w)| + |g(w)-g(u)| \leqslant \psi(|\mathbf{q}|) + 2m^{1/2}|\mathbf{q}|kK|w-u| \leqslant 2\psi(|\mathbf{q}|).$$

Thus $u_1 \in E(u')$ and so $w \in F(u')$, where $u' \in \Lambda^{m-1}$, proving the lemma in this case.

The proof of the more difficult 'non-transverse' case where $|\mathbf{q}^{(2)}| > |\mathbf{q}|/(2kK)$ is again similar to the same case in Lemma 2.19. Make the orthogonal transformation O_v of the u coordinates in $\mathbb{R}^m$ which diagonalises the Hessian $D^2g(0)$ to $\operatorname{diag}(\lambda_1, \lambda_2, \ldots, \lambda_m)$. For simplicity the old notation will be retained for the transformed variables and functions. By relabelling if necessary, we take $|\lambda_1| \geqslant 2\delta$ and $|\lambda_2| \geqslant 2\delta$ for some $\delta > 0$. It follows from (2.45) that for $u \in U$ the elements of $D^2g(u)$ are close to the elements of $D^2g(0)$. In particular each leading diagonal element λ_i is close to $\partial^2 g(u)/\partial u_i^2$. Hence for each $u \in \mathbb{R}^m$ and $i = 1, 2$,

$$\left|\frac{\partial^2 g(u)}{\partial u_i^2}\right| \geqslant |\mathbf{q}^{(2)}|\,\delta,$$

and the off-diagonal elements satisfy

$$\left|\frac{\partial^2 g(u)}{\partial u_1 \partial u_2}\right| \leqslant |\mathbf{q}^{(2)}|\,\delta/2.$$

Write $u = (u_1, u_2, u')$ where now $u' = (u_3, \ldots, u_m) \in \mathbb{R}^{m-2}$ and define

$$E(u_2, u') = \{u_1 \in \mathbb{R} \colon |g(u_1, u_2, u')| < 2\,\psi(|\mathbf{q}|)\}.$$

For each fixed $u' \in \mathbb{R}^{m-2}$, the function $h\colon (u_1, u_2) \mapsto g(u_1, u_2, u')$ is a C^3 function of two real variables. Construct the function $h_0\colon u_2 \mapsto g_0(u_2, u')$ as in (4.8), and the number $h_{00} := g_{00}(u')$ as in (4.9) above. By the implicit function theorem, the functions $(u_2, u') \mapsto g_0(u_2, u')$ and $u' \mapsto g_{00}(u')$ are C^2 and C^1 functions on $\mathbb{R}^{m-1}$ and $\mathbb{R}^{m-2}$ respectively.

It follows from (4.7) and Lemma 4.5 that the set $E(u_2, u')$ is the union of at most two intervals and has measure

$$|E(u_2,u')| \ll (\psi(|\mathbf{q}|)/|\mathbf{q}|)^{1/2} \min\{\psi(|\mathbf{q}|)^{1/2}\, |g_0(u_2,u')|^{-1/2}, 1\}.$$

For fixed u_2, u', the set

$$F(u_2,u') = \{w \in U\colon \text{ dist}\,(w, (E(u_2,u'), u_2, u')) < \eta(|\mathbf{q}|)\}$$

can be covered by a collection $\mathscr{C}(u_2,u')$ of $\ll |E(u_2,u')|/\eta(|\mathbf{q}|)$ hypercubes C in $\mathbb{R}^m$ with $\ell(C) = \eta(|\mathbf{q}|) = \psi(|\mathbf{q}|)/|\mathbf{q}|$.

Given $u' \in \mathbb{R}^{m-2}$ and $N \in \mathbb{Z}$, the set

$$G_N(u') = \{u_2 \in \mathbb{R}\colon\ -\psi(|\mathbf{q}|) < g_0(u_2,u') - 2\,N\,\psi(|\mathbf{q}|) \leqslant \psi(|\mathbf{q}|)\}$$

is by (4.7), Lemma 4.5 and (4.9), the union of two intervals and

$$|G_N(u')| \ll \min\{\psi(|\mathbf{q}|)|\mathbf{q}|^{-1/2}|g_{00}(u') - 2N\psi(|\mathbf{q}|)|^{-1/2},\ (\psi(|\mathbf{q}|)/|\mathbf{q}|)^{1/2}\}.$$

Let $N_0 = [1/2 + g_0(u_2,u')/2\psi(|\mathbf{q}|)]$ be the unique integer such that

$$-\psi(|\mathbf{q}|) < g_0(u_2,u') - 2\,N_0\,\psi(|\mathbf{q}|) \leqslant \psi(|\mathbf{q}|).$$

For convenience we will write η instead of $\eta(|\mathbf{q}|)$ for the rest of this section. Then for each $u_2 \in G_N(u')$,

$$|E(u_2,u')| \ll \begin{cases} \eta^{1/2} & \text{when } N = 0,\\ (\eta/|N|)^{1/2} & \text{when } N \neq 0, \end{cases}$$

while

$$|G_N(u')| \ll \begin{cases} \eta^{1/2} & \text{when } N = N_0,\\ (\eta/|N-N_0|)^{1/2} & \text{when } N \neq N_0. \end{cases}$$

Let

$$\mathscr{C}(u') = \bigcup_{N\in\mathbb{Z}} \bigcup_{u_2} \mathscr{C}(u_2,u'),$$

where the second union is over the set $\Lambda^1 \cap G_N(u')$ containing $\ll |G_N(u')|/\eta$ points. In addition, since we can take $M \cap V \subset (-1/2,1/2)^n$, it follows that $|g(u)-g(v)| \ll |\mathbf{q}|$ for each $u, v \in (-1/2,1/2)^m$, whence the number of integers N for which $\Lambda^1 \cap G_N(u')$ is non-empty is $\ll \eta^{-1}$. Hence the number of hypercubes C in $\mathscr{C}(u')$ is

$$\begin{aligned} &\ll \eta^{1/2}\,\eta^{-1}\,\eta^{1/2}\,\eta^{-1} + \sum_{\substack{|N|\ll\eta^{-1}\\ N\neq 0,N_0}} \eta^{1/2}\,|N|^{-1/2}\,\eta^{-1}\,\eta^{1/2}\,|N-N_0|^{-1/2}\,\eta^{-1}\\ &\ll \eta^{-1}\Big(1 + \sum_{\substack{|N|\ll\eta^{-1}\\ N\neq 0,N_0}} |N|^{-1/2}|N-N_0|^{-1/2}\Big) \ll \eta^{-1}(1+|\log\eta|). \end{aligned}$$

Thus the total number of hypercubes C in

$$\mathscr{C}(p,\mathbf{q}) = \bigcup_{u' \in \Lambda^{m-2}} \mathscr{C}(u')$$

is $\ll \eta^{-m+2}\eta^{-1}\,|\log\eta|$ and so $\ell^s(\mathscr{C}(p,\mathbf{q})) \ll \eta^{s-m+1}\,|\log\eta|$. By a mean value argument similar to the one above, $\mathscr{C}(p,\mathbf{q})$ covers $B_{\psi(|\mathbf{q}|)}(p,\mathbf{q})$.

Since U is bounded, for each non-zero $\mathbf{q} \in \mathbb{Z}^n$ there are $\ll |\mathbf{q}|$ integers p such that $B_{\psi(|\mathbf{q}|)}(p,\mathbf{q})$ is non-empty. Let

$$\mathscr{C}(\mathbf{q}) = \bigcup_{p\in\mathbb{Z}} \mathscr{C}(p,\mathbf{q}). \tag{4.11}$$

Then for each non-zero $\mathbf{q} \in \mathbb{Z}^n$, the collection $\mathscr{C}(\mathbf{q})$ of hypercubes C with side-length $\ell(C) = \eta = \psi(|\mathbf{q}|)/|\mathbf{q}|$ covers $B_{\psi(|\mathbf{q}|)}(\mathbf{q})$ and

$$\ell^s(\mathscr{C}(\mathbf{q})) \ll \eta^{s-m+1}(1 + |\log\eta|)|\mathbf{q}|$$

and the lemma follows. □

Now we can obtain the upper bound for $\dim \mathscr{L}(M;\psi)$.

LEMMA 4.7. *Let M be a C^3 m-dimensional manifold embedded in $\mathbb{R}^n$, where $m \geqslant 2$. Let M_0 be the set of points $x \in M$ at which M is not 2-curved and suppose that* $\dim M_0 \leqslant m-1$. *Then for any decreasing function $\psi\colon \mathbb{N} \to \mathbb{R}^+$,*

$$\dim \mathscr{L}(M\,;\psi) \leqslant m - 1 + \frac{n+1}{\lambda+1},$$

where λ is the lower order of $1/\psi$.

PROOF. It suffices to consider a suitable Monge domain U. It has been shown above that $L(\psi)$ can be covered by the collection $\mathscr{C} = \{\mathscr{C}(\mathbf{q})\colon \mathbf{q} \in \mathbb{Z}^n \setminus 0\}$, where $\mathscr{C}(\mathbf{q})$ is given by (4.11). Now

$$\begin{aligned}\ell^s(\mathscr{C}) &\ll \sum_{C\in\mathscr{C}} \ell(C)^s \ll \sum_{\mathbf{q}\in\mathbb{Z}^n\setminus\{0\}} \psi(|\mathbf{q}|)^{s-m+1}\,|\mathbf{q}|^{-s+m}\,|\log(\psi(|\mathbf{q}|)/|\mathbf{q}|)| \\ &\ll \sum_{q=1}^{\infty} \psi(q)^{s-m+1}\,q^{-s+m+n-1}|\log(\psi(q)/q)|.\end{aligned}$$

Since λ is the lower order of $1/\psi$, given any $\varepsilon > 0$, $\psi(q) \ll q^{-\lambda+\varepsilon}$ for q sufficiently large. On substituting, we get

$$\ell^s(\mathscr{C}) \ll \sum_{q=1}^{\infty} q^{-(\lambda-\varepsilon)(s-m+1)-s+m+n-1}\,\log q < \infty$$

when $s > m-1+(n+1)/(\lambda+1-\varepsilon)$ for arbitrary $\varepsilon > 0$. Hence by the Hausdorff-Cantelli lemma, any Monge domain U,

$$\dim L(\psi) \leqslant m - 1 + \frac{n+1}{\lambda+1}$$

when $\lambda \geqslant n$. The proposition now for $\lambda \geqslant n$ follows from (3.18). □

A general complementary result is proved in Chapter 5 (Lemma 5.11). When $\psi(q) = q^{-v}$, the lower order λ of $1/\psi$ is v and so

$$\dim \mathscr{L}_v(M) \leqslant m - 1 + \frac{n+1}{v+1} < m$$

for $v > n$, whence $|\mathscr{L}_v(M)| = 0$ for $v > n$.

4.4. Simultaneous Diophantine approximation

4.4.1. The circle $\mathbb{S}^1$. That simultaneous approximation is harder that the dual form can be seen by considering the Hausdorff dimension of the set

$$\mathscr{S}_v(\mathbb{S}^1) = \{(x,y) \in \mathbb{S}^1 : |\langle (qx, qy) \rangle| < q^{-v} \text{ for infinitely many } q \in \mathbb{N}\},$$

studied by Melnichuk [43, 161]. When $v \leqslant 1/2$, $\mathscr{S}_v(\mathbb{S}^1) = \mathbb{S}^1$ and when $v \geqslant 1/2$, R. C. Baker's general result [16] implies that the set L_v of points $u \in (-1,1)$ which satisfy

$$\|q_1 u + q_2\sqrt{1-u^2}\| < |(q_1, q_2)|^{-v}$$

for infinitely many $(q_1, q_2) \in \mathbb{Z}^2$, has Hausdorff dimension $\dim L_v(\mathbb{S}^1) = 3/(v+1)$. It follows that $\dim \mathscr{L}_v(\mathbb{S}^1) = 3/(v+1)$ for $v \geqslant 2$. Now the set E of points on $\mathbb{S}^1$ with a rational coordinate is countable and so by Lemma 3.6, $\dim E = 0$. It follows from Khintchine's transference principle (see §1.3.1) that for any $\varepsilon > 0$,

$$\mathscr{L}_{2v/(1-v-\varepsilon)}(\mathbb{S}^1) \setminus E \subset \mathscr{S}_v(\mathbb{S}^1) \setminus E \subset \mathscr{L}_{2v+1-\varepsilon}(\mathbb{S}^1) \setminus E,$$

whence for $1/2 \leqslant v \leqslant 1$ the inequality

$$\frac{3(1-v)}{v+1} \leqslant \dim \mathscr{S}_v(\mathbb{S}^1) \leqslant \frac{3}{2(v+1)}$$

holds. This is sharper than the result using exponential sums obtained in [161].

Melnichuk also studied the case $v > 1$ in [161]. Let (x, y) lie on $\mathscr{S}_v(\mathbb{S}^1)$. Then

$$|\langle (qx, qy) \rangle| < q^{-v}$$

for some integer $q \geqslant 2$ and so there exist integers p, r, $0 \leqslant |p|, |r| \leqslant 2q$, such that

$$\max\left\{\left|x - \frac{p}{q}\right|, \left|y - \frac{r}{q}\right|\right\} < q^{-v-1}. \tag{4.12}$$

When $v > 1$ and q is a sufficiently large positive integer, any point $(p/q, r/q)$ which satisfies the above inequality lies on the unit circle $\mathbb{S}^1$. For on rearranging we get

$$(x, y) = (p/q + \alpha q^{-v-1}, r/q + \beta q^{-v-1})$$

where $|\alpha|, |\beta| \leqslant 1$, whence

$$q^2 = p^2 + r^2 + 2(p\alpha + r\beta)q^{-v} + (\alpha^2 + \beta^2)q^{-2v}.$$

But for q sufficiently large, this equation is only soluble if $q^2 = p^2 + r^2$. Hence when $v > 1$, the solutions $(p/q, r/q)$ to (4.12) lie on $\mathbb{S}^1$. Positive integers p, r, q satisfying $p^2 + r^2 = q^2$ are called a *Pythagorean triple* [12].

When p, r are coprime and r is even, the Pythagorean triples (p, r, q) are generated by positive coprime integers a, b of opposite parity (*i.e.*, not both even or not both odd) as follows:

$$p = a^2 - b^2, \quad r = 2ab, \quad q = a^2 + b^2. \tag{4.13}$$

The number of such pairs (a, b) with $a^2 + b^2 \leqslant Q$ is at most the number of coprime lattice points (a, b) of opposite parity in a disc of radius $Q^{1/2}$ and so is $\ll Q$. The points $(p/q, r/q)$, $(r/q, p/q)$ where p and r are coprime to q, lies in the positive quadrant of $\mathbb{S}^1$ if and only if there are coprime positive integers a, b with $a > b$ of opposite parity and satisfying (4.13).

Now consider the Monge parametrisation $(1_U \times \varphi, U)$ for the circle $\mathbb{S}^1$ and the set $S_v\ (= \pi_U(\mathscr{S}_v(\mathbb{S}^1)))$. Choose one domain U to be the projection of A_1, the upper arc subtending $\pi/3$. The Monge map $\varphi\colon (-1/2, 1/2) \to \mathbb{R}$ is given by $\varphi(u) = \sqrt{1-u^2}$, has graph A_1 and is bi-Lipschitz. The set

$$S_v = \{u \in (-1/2, 1/2)\colon u \in B(q) \cap (-1/2, 1/2) \text{ for infinitely many } q \in \mathbb{N}\},$$

i.e., S_v is the set of points $u \in (-1/2, 1/2)$ which fall into infinitely many sets

$$\begin{aligned} B(q) &= \{u \in (-1/2, 1/2)\colon |\langle q(u, \sqrt{1-u^2})\rangle| < q^{-v}\} \\ &\subset \bigcup_{p=-q}^{q} \left\{ u \in I\left(\frac{p}{q}; q^{-v-1}\right) : \left|\sqrt{1-u^2} - \frac{r}{q}\right| < \frac{1}{q^{v+1}} \text{ for some } r \in \mathbb{N} \right\}. \end{aligned}$$

Let A_2, A_3, A_4 be the rotations of A_1 by $\pi/2$, π and $3\pi/2$ respectively, so that $\mathbb{S}^1 = \cup_{j=1}^4 A_j$.

Let $\mathcal{N}(Q)$ be the number of positive rationals pairs $(p/q, r/q)$ with $q \leqslant Q$ for which p, r, q are a Pythagorean triple. Taking symmetries into account, the number of distinct rational points $(p/q, r/q)$ in their lowest terms with $2^t \leqslant q \leqslant 2^{t+1}$ lying on the upper semicircle of $\mathbb{S}^1$ is

$$\ll \mathcal{N}(2^{t+1}) - \mathcal{N}(2^t) \ll 2^t.$$

Since the length of the (non-empty) interval $I(p/q; 1/q^{v+1})$ is $\ll (2^t)^{-v-1}$ and since p, q can be taken to be coprime without loss of generality, S_v has a natural cover of intervals with s-length

$$\ll \sum_{t=0}^{\infty} 2^{-t(v+1)s} \left(\mathcal{N}(2^{t+1}) - \mathcal{N}(2^t)\right) \ll \sum_{t=0}^{\infty} 2^{-t(v+1)s+t} < \infty$$

when $s > 1/(v+1)$. Hence

$$\dim \mathscr{S}_v(A_1) = \dim S_v \leqslant 1/(v+1)$$

for $v > 1$ by the Hausdorff-Cantelli lemma. The other arcs A_2, A_3 and A_4 are rotations of A_1 and so have the same Hausdorff dimension, whence

$$\dim \mathscr{S}_v(\mathbb{S}^1) \leqslant 1/(v+1) \tag{4.14}$$

for $v > 1$. The lower bound for $\dim \mathscr{S}_v(\mathbb{S}^1)$ is discussed briefly in §5.4.

Very close simultaneous Diophantine approximation (in the sense that the exponent v in (4.12) exceeds $k-1$) on the planar curve $M = \{(x,y) \colon x^k + y^k = 1\}$ is possible for at most four points when $k \geqslant 3$.

THEOREM 4.8. *When $k \geqslant 3$ and $v > k-1$, the set $S_v(\mathbb{S}^1)$ of points (x,y) on the curve $\{(x,y)\colon x^k + y^k = 1\}$ and satisfying*

$$|\langle(qx, qy)\rangle| < q^{-v} \tag{4.15}$$

for a sufficiently large integer q, contains at most four points.

PROOF. The point (x,y) satisfying (4.15) can be written in the form

$$(x, (1-x^k)^{1/k}) = (p/q + \alpha q^{-v-1}, r/q + \beta q^{-v-1})$$

where $|\alpha|, |\beta| \leqslant 1$. Hence $q^k = p^k + r^k + (\alpha_1 + \beta_1) q^{-v-1+k}$, where $|\alpha_1|, |\beta_1| < k2^{k-1}$. If $q^k \neq p^k + r^k$, then

$$1 \leqslant |q^k - p^k - r^k| < k2^k q^{-v-1+k}$$

which since q is sufficiently large, is impossible for $v > k-1$. But by Wiles' theorem [216, 230], when $k \geqslant 3$, $q^k = p^k + r^k$ is not soluble in positive integers p, r, q. Thus if k is odd $S_v(\mathbb{S}^1) = \{(1,0),(0,1)\}$ and if k is even $S_v(\mathbb{S}^1) = \{(\pm 1, 0), (0, \pm 1)\}$. □

Thus when M is the curve $\{(x,y)\colon x^k + y^k = 1\}$, $k \geqslant 3$, the set $\mathscr{S}_v(M)$ is finite for $v > k-1$ and is M when $v \leqslant 1/2$. Hence by (4.21),

$$\dim \mathscr{S}_v(M) \begin{cases} = 1, & v \leqslant 1/2, \\ \leqslant 3/2(v+1), & 1/2 < v < n-1, \\ = 0, & v \geqslant n-1. \end{cases}$$

By contrast $\dim \mathscr{L}_v(M) = 3/(v+1)$ for $v > 2$ by [16]. This illustrates the difficulty of obtaining correct or even good upper bounds for the Hausdorff dimension of sets of simultaneously v-approximable points.

4.4.2. Two other manifolds. When the parametrisations satisfy suitable analytic and algebraic conditions, exponential sums, a powerful and widely used technique in number theory, can be used to obtain estimates for the number of rational points near a manifold. These lead to estimates for the Hausdorff dimension of simultaneously well approximable points on manifolds. Lower estimates using exponential sum methods will be discussed in more detail in §5.4.

In this section we consider two special manifolds discussed in Chapter 2, §4 in [210]. In order to conform with the rest of the tract, the notation used here is not the same and we will also assume for simplicity (and without significant loss of generality) that $U = [0, 1]^m$. As usual, we discuss the Hausdorff dimension of the set

$$S_v = \{u \in [0,1]^m : |\langle(qu, q\varphi(u))\rangle| < q^{-v} \text{ for infinitely many } q \in \mathbb{N}\}$$

instead of $\mathscr{S}_v(M)$, and we take $v > 1/n$ since otherwise $S_v = [0,1]^m$. Upper estimates for the Hausdorff dimension of S_v are obtained by replacing the natural cover by a hypercube cover.

Suppose the manifold M is C^2. For each $u \in S_v$ there exist infinitely many positive integers q such that

$$|\langle(qu, q\varphi(u))\rangle| < q^{-v}. \tag{4.16}$$

The family of sets $B(q) = \{u \in [0,1]^m : |\langle(qu, q\varphi(u))\rangle| < q^{-v}\}$, $q = 1, 2, \ldots$, is a natural cover for S_v. Given a sufficiently large q satisfying (4.16), there exist a vector $\mathbf{a} \in \mathbb{Z}^m$ and a unique vector $\mathbf{b} = \mathbf{b}(\mathbf{a}) \in \mathbb{Z}^k$ such that

$$\left|u - \frac{\mathbf{a}}{q}\right|, \left|\varphi(u) - \frac{\mathbf{b}}{q}\right| < q^{-v-1}. \tag{4.17}$$

By Taylor's theorem

$$\varphi(u) = \varphi\left(\frac{\mathbf{a}}{q}\right) + O\left(\max\left\{\left|\frac{\partial\varphi(u)}{\partial u_i}\right| : i = 1, \ldots, m\right\}\left|u - \frac{\mathbf{a}}{q}\right|\right) = \varphi\left(\frac{\mathbf{a}}{q}\right) + \mathcal{R},$$

where $|\mathcal{R}| \leqslant c|u - \mathbf{a}/q|$ for some $c > 0$. Thus if u satisfies (4.17), then

$$|\varphi(\mathbf{a}/q) - \mathbf{b}/q| < c|u - \mathbf{a}/q| + q^{-v-1} < (c+1)q^{-v-1}. \tag{4.18}$$

These inequalities imply that $u \in B(q, \mathbf{a})$, where

$$B(q, \mathbf{a}) = \prod_{i=1}^{m}\left(\frac{a_i}{q} - \frac{1}{q^{v+1}}, \frac{a_i}{q} + \frac{1}{q^{v+1}}\right) \cap \varphi^{-1}\left(\prod_{j=1}^{k}\left(\frac{b_j}{q} - \frac{1}{q^{v+1}}, \frac{b_j}{q} + \frac{1}{q^{v+1}}\right)\right)$$

and

$$\max\{|q\varphi(u) - q\varphi(\mathbf{a}/q)|, \|q\varphi(\mathbf{a}/q)\|\} \ll q^{-v}, \tag{4.19}$$

where $0 \leqslant a_i \leqslant q$, $i = 1, \ldots, m$.

Given q, let $\mathcal{N}_v(q)$ be the number of integer vectors $\mathbf{a}$ in $q\,[0,1]^m = [0,q]^m$ such that the inequalities (4.19) hold or such that $B(q, \mathbf{a})$ is not empty. Then

$$B(q) \subseteq \bigcup_{\mathbf{a} \in q[0,1]^m} B(q, \mathbf{a}) \subseteq \bigcup_{\mathbf{a} \in q[0,1]^m} \prod_{i=1}^{m}(a_i/q - q^{-v-1}, a_i/q + q^{-v-1})$$

and $B(q)$ has a cover $\mathscr{C}_q$ by $\mathcal{N}_v(q)$ hypercubes of sidelength $O(q^{-v-1})$. Thus S_v has a hypercube cover $\mathscr{C}$ with s-length

$$\ell^s(\mathscr{C}) \ll \sum_{C \in \mathscr{C}} \ell(C)^s \ll \sum_{q=1}^{\infty} q^{-s(v+1)}\mathcal{N}_v(q). \tag{4.20}$$

In order to get a non-trivial estimate for $\mathcal{N}_v(q)$, we suppose that for each $j = 1, \ldots, k$, the ordinate function φ of the manifold M has the special form

$$\varphi_j(u) = \varphi_j(u_1, \ldots, u_m) = a_{j1}\vartheta_1(u_1) + \cdots + a_{jm}\vartheta_m(u_m),$$

where $a_{ji} \in \mathbb{Q}$, $i = 1, \ldots, m$, $j = 1, \ldots, k$, are such that $\operatorname{rank}(a_{ij}) = k$ and where for each $j = 1, \ldots, m$, $\vartheta_j \colon [0,1] \to \mathbb{R}$ are C^3 functions with $|\varphi_j''(u)|$ bounded away from 0. Using Lemma 4.9 below, it is shown on p. 95 of [210] that for such functions $\mathcal{N}_n(q) \ll \mathcal{N}_k(q) \ll q^{m-k/n+\varepsilon_1}$, where $\varepsilon_1 > 0$ is arbitrary. Thus when $s > (m+1-k/n)/(v+1)$, it follows from (4.20) that

$$\sum_{C \in \mathscr{C}} \ell(C)^s \ll \sum_{q=1}^{\infty} q^{-(v+1)s} q^{m-k/n+\varepsilon_1} < \infty.$$

Hence by the Hausdorff-Cantelli lemma and since $k = n - m$,

$$\dim S_v \leqslant \frac{m(n+1)}{n(v+1)}. \tag{4.21}$$

One can also refine Theorem 4, p. 82 in [210] in a similar way.

Improving the estimate for $\mathcal{N}_v(q)$ would lead via (4.20) to an improvement in the upper bound for the Hausdorff dimension. To illustrate this, consider the simple case where the manifold M is essentially the product of planar curves with non-vanishing curvatures. Accordingly let $m \geqslant 2$ and $m = k$, corresponding to the m-dimensional manifold M embedded in $\mathbb{R}^{2m}$ being a product $\prod_{i=1}^{m} \Gamma_i$ of planar curves Γ_i in Monge form:

$$\Gamma_i = \{(u, \varphi_i(u)) \colon u \in [0,1]\}.$$

After a harmless change of coordinates, each point $\xi \in M$ can be written

$$\xi = (u_1, \ldots, u_m, \varphi_1(u_1), \ldots, \varphi_m(u_m)).$$

We assume that each φ_i is C^2, so that φ_i and its first two derivatives are bounded in $[0,1]$ by constants K_i, K_i' and K_i'' respectively. The requirement that the curvature does not vanish implies that $|\varphi_i''|$ is bounded away from 0 on $[0,1]$.

The set S_v consists of points $u = (u_1, \ldots, u_m) \in U = [0,1]^m$ for which the inequality

$$\max\{\|qu_j\|, \|q\varphi_j(u_j)\| \colon j = 1, \ldots, m\} = |\langle q(u, \varphi_1(u_1), \ldots, \varphi_m(u_m))\rangle| < q^{-v}$$

has an infinite number of solutions $q \in \mathbb{N}$. Note that if $v \leqslant 1/n = 1/(2m)$, then $S_v = [0,1]^m$, so as usual we will take $v > 1/n$.

Consider first the planar curve $\{(u, \varphi(u)) \colon u \in [0,1]\}$. We need two results. The first is based on the observation that when $\|\theta\|$ is small, $e^{2\pi i\theta}$ is close to 1, and is Lemma 7, p. 85 in [210], with $n = 1$.

LEMMA 4.9. *Let q, Q be positive integers and let v, g_a, $a = 0, 1, \ldots, Q$, be positive numbers. The number of integers a, $0 \leqslant a \leqslant Q$, for which the inequalities*

$$\|g_a\| < q^{-v}$$

hold is $\ll q^{-v} \sum_{|j|<q^v} |\sum_{a=0}^{Q} e^{2\pi i j g_a}|$.

The second lemma is due to van der Corput (a convenient reference is [179, p. 104]).

LEMMA 4.10. *Suppose $f \in C^2([a, b], \mathbb{R})$ where $b - a \geqslant 1$, and suppose that*

$$\inf\{f''(x) : x \in [a, b]\} > 0.$$

Then

$$\sum_{a<n\leqslant b} e^{2\pi i f(n)} \ll ((b-a)\sup f''(x) + 1)\,(\inf f''(x))^{-1/2}.$$

Thus when $f(u) = j\varphi(u/q)$, $j > 0$, it follows that $f''(u) = j\varphi''(u/q)/q$ and so since $\varphi''(u) \asymp 1$,

$$\left|\sum_{a=0}^{q} e^{2\pi i j\varphi(a/q)}\right| \ll (q(j/q) + 1)\,(q/j)^{1/2} \ll (jq)^{1/2} + (q/j)^{1/2}.$$

Let $\mathcal{N}_v(q)$ be the number of a, $0 \leqslant a \leqslant q$, such that $\|\varphi(a/q)\| < q^{-v}$. Then by Lemma 4.9,

$$\begin{aligned}\mathcal{N}_v(q) &\ll q^{1-v} + q^{-v} \sum_{1\leqslant j\leqslant q^v} \left|\sum_{a=0}^{q} e^{2\pi i j\varphi(a/q)}\right| \ll q^{1-v} + q^{1/2-v} \sum_{1\leqslant j\leqslant q^v} j^{1/2} \\ &\ll q^{1-v} + q^{1/2-v+3v/2} \ll q^{1-v} + q^{(1+v)/2}.\end{aligned}$$

Therefore when $v < 1/3$, we have that $\mathcal{N}_v(q) \ll q^{1-v}$.

Now we return to the manifold $M = \prod_{i=1}^{m} \Gamma_i$. Let $\mathcal{N}_v^{(m)}(q)$ be the number of integer vectors $\mathbf{a} \in q\,[0, 1]^m$ such that $|\langle q\,\varphi(\mathbf{a}/q)\rangle| \leqslant q^{-v}$. Then

$$\mathcal{N}_v^{(m)}(q) \ll \mathcal{N}_v(q)^m$$

and so for each $N = 1, 2, \ldots$, the set S_v has a cover

$$\mathscr{C} = \left\{\prod_{i=1}^{m} (a_i/q - q^{-v-1}, a_i/q + q^{-v-1}) : 0 \leqslant a_i \leqslant q, q \in \mathbb{N}\right\}$$

consisting of hypercubes C with $\ell(C) \leqslant q^{-v-1}$. From (4.20), when $v < 1/3$, the s-length of the cover $\mathscr{C}$ satisfies

$$\ell^s(\mathscr{C}) \ll \sum_{q=1}^{\infty} q^{-s(v+1)}\,\mathcal{N}_v^{(m)}(q) \ll \sum_{q=1}^{\infty} q^{-s(v+1)}\,q^{m(1-v)} < \infty$$

when $s > (1 + m(1-v))/(1+v)$. Hence for $1/(2m) < v < 1/3$,

$$\dim \mathscr{S}_v(\prod_{i=1}^{m} \Gamma_i) = \dim S_v \leqslant \frac{m + 1 - mv}{v + 1}.$$

The same estimate can be obtained using Fourier series in a manner similar to that in §5.4. Indeed estimates of this type are used there to obtain regular and ubiquitous systems to give lower bounds as well. In the case $v \geqslant 1/3$, it is shown in [45] that the obvious inequalities $\mathcal{N}_v \leqslant \mathcal{N}_{1/3} \leqslant q^{2/3}$ give

$$\dim S_v \leqslant \frac{2m+3}{3(1+v)}. \tag{4.22}$$

4.5. Notes

§ **4.1.1** The Jarník-Besicovitch theorem was first extended to systems of linear forms in [54]. H. Dickinson and S. L. Velani [79] have extended Jarník's theorem on simultaneous Diophantine approximation [127] to systems of linear forms, thus obtaining the Hausdorff measure counterpart of the Khintchine-Groshev theorem.

The proof of Lemma 4.1 extends with the obvious modifications to the inhomogeneous case where the inequality is $|\langle \mathbf{q}A + \alpha\rangle| < \psi(|\mathbf{q}|)$ for a given $\alpha \in \mathbb{R}^n$ (see [84], [149]). For results on restricted Diophantine approximation, see [52], [192] and for applications to inhomogeneous approximation and to non-monotonic approximation functions, see [75], [124].

The notion of a 'sparse' system, introduced recently by Falconer in [102], leads to best possible dimension estimates.

§ **4.1.2** B. P. Rynne [194] has studied simultaneous approximation with differing exponents in the approximation functions for each coordinate; the Hausdorff dimension turns out to depend only on the degree of approximation of one coordinate. I. R. Dombrowsky [97] has also studied this type of simultaneous approximation but with approximation by algebraic numbers.

§ **4.2.2** D. V. Vasilyev [221] has shown that when $(z_1, z_2) = (f_1(z), f_2(z))$ and f_1, f_2 are complex analytic functions with $f_1'(z)f_2''(z) \neq f_1''(z)f_2'(z)$ for at least one point on an open set in $\mathbb{C}$, the Hausdorff dimension of the set of points for which the inequality (4.4) holds infinitely often is $3/(v+1)$ for $v \geqslant 1/2$. This generalises R. C. Baker's result for the curve $\{(z^m, z^n) \colon z \in \mathbb{C} \setminus \mathbb{R}\}$, where m, n are integers with $1 \leqslant m < n$ [16] (see also [208, Chapter 1, Lemma 14]), to complex analytic curves. The complementary lower bound for the Hausdorff dimension is obtained using regular systems, discussed in Chapter 5.

§ **4.2.3** Theorem 4.7 also implies that such manifolds M are extremal, see §2.2.

Distance functions $F \colon \mathbb{R}^n \to \mathbb{R}$ are continuous, non-negative and satisfy $F(tx) = tF(x)$ for all $t \geqslant 0$ ([60, p. 103]); they are closely related to star bodies [111]. The 'sup' norm and the usual Euclidean norm are distance functions ($F(x) = |x|$ and $F(x) = |x|_2$ respectively), as is $F(x) = \prod_{j=1}^{n} |x_j|^{1/n}$ of interest in the geometry of numbers [111, Chapter 4], uniform distribution [210, p. 69] and strongly extremal manifolds [139]. Some general upper bounds for sets $\{x \in \mathbb{R}^n \colon F(\langle qx\rangle) < \psi(q)$ for infinitely many $q \in \mathbb{N}\}$ and related systems of linear forms have been obtained in [81], [85]; the variables here are, however, independent.

CHAPTER 5

Lower bounds for Hausdorff dimension

5.1. Introduction

When the variables are independent, the correct lower bound for the Hausdorff dimension of sets of ψ-approximable points can be much harder to determine than the upper bound, as has been pointed out in §3.5.2 for the Jarník-Besicovitch theorem. When manifolds (or dependent variables) are considered, simultaneous Diophantine approximation involves the distribution of rational points of the form $\mathbf{p}/q$ near or on the manifold. Estimates for the number of rational points near the manifold lead to lower bounds for the Hausdorff dimension. In the dual case of linear forms, the functional relations between the coordinates are less serious for the lower bound than the upper bound. Regular and ubiquitous systems have proved very useful in obtaining lower bounds for the Hausdorff dimension of sets of ψ-approximable points. We begin with regular systems.

5.2. Regular systems

These were introduced by A. Baker and W. M. Schmidt [13] as a means of obtaining lower bounds for the Hausdorff dimension of the sets $\mathfrak{K}_w$ and $\mathscr{M}_v$ (see §4.1.2 and §4.2.3). These results reduce to the Jarník-Besicovitch theorem in the one dimensional case and indeed some of the underlying ideas go back to Besicovitch's proof in [48]. Regular systems have turned out to be a very effective technique and have been extended and applied widely. In his refinement [15] of Schmidt's thereom on planar curves [198], R. C. Baker used them to obtain the lower bound for the Hausdorff dimension of the exceptional set of dually v-approximable points on planar curves (discussed in §2.2) and the Minsk school have developed and made extensive use of the technique (see [22], [23], [27], [39], [45], [165], [221]).

To maintain consistency, the notation here differs from [13]. Also, for simplicity, the dimension function h will be taken to be $h(x) = x^s$. The pair (R, ν), where R ($= \Gamma$ in [13]) is a countable set of real numbers and $\nu\colon R \to \mathbb{R}^+$ ($\nu(r) = N(\gamma)$ in [13]) is a function, is called a *regular system with respect to an interval* I_0 if there exists a positive constant $c = c\,(R, \nu)$ such that for each finite interval $I \subseteq I_0$ there exists a positive number $K(I)$ such that for each $K \geqslant K(I)$, there exist t points $r_1, \dots, r_t \in R \cap I$ (depending on K) where $t \geqslant c|I|K$ (recall that $|I|$ is the Lebesgue measure of I) and which for distinct i, j, $1 \leqslant i, j \leqslant t$, satisfy

$$\nu(r_i) \leqslant K, \ |r_i - r_j| \geqslant 1/K.$$

As the name suggests, a regular system contains plenty of well spaced points. A simple (and motivating) example is to take R to be the rationals $\mathbb{Q}$ and $\nu(p/q) = q^2$

when p, q are coprime. The points $r_1, \ldots, r_t$ in R are rationals p/q satisfying $\nu(p/q) = q^2 \leqslant K$, where $K \geqslant K(I)$ and where $K(I)$ is chosen large enough to ensure that the interval I contains a rational p/q with $q^2 = \nu(p/q) \leqslant K(I)$.

Given a function $\psi\colon \mathbb{R}^+ \to \mathbb{R}^+$, let $\Lambda(R, \nu; \psi)$ be the set of all real numbers ξ for which the inequality $|\xi - r| < \psi(\nu(r))$ holds for infinitely many r in R. When (R, ν) is a regular system and when the function ψ is monotonic with $\psi(q) \leqslant 1/(2q)$, a lower bound for the Hausdorff dimension of $\Lambda(R, \nu; \psi)$ can be established in terms of (R, ν) and ψ.

For any $S \subset \mathbb{R}$ and positive s write $S \prec x^s$ if for every positive δ and η, the set S can be covered by a countable family $\mathcal{I}_\eta(\delta, x^s)$ of intervals I_j with $|I_j| \leqslant \delta$ and such that

$$\sum_{j=1}^{\infty} |I_j|^s < \eta.$$

It is clear that if $S_1 \prec x^s$ and $S_2 \prec x^s$, then $S_1 \cup S_2 \prec x^s$. If S is countable then by covering each $a_j \in S$ in the standard way with an interval centred at a_j and of length $2^{-j}\eta$, it follows that $S \prec x^\varepsilon$ for every $\varepsilon > 0$. Moreover the definition of Hausdorff dimension implies that if $\dim S < s$ then $S \prec x^s$; or in the equivalent contrapositive form, $S \not\prec x^s$ implies that any cover of S has s-length bounded away from 0, whence $\mathcal{H}^s(S) > 0$ and $\dim S \geqslant s$.

Now suppose $\psi\colon \mathbb{R}^+ \to \mathbb{R}^+$ is decreasing and that $x\psi(x) \leqslant 1/2$ for large x. Further suppose that $0 < s < 1$ and that $x\psi(x)^s \to \infty$ as $x \to \infty$. Then for any regular system (R, ν), $\Lambda(R, \nu;\, \psi) \not\prec x^s$ (see Lemma 1 of [13] for details). This implies the following lower bound for the Hausdorff dimension of $\Lambda(R, \nu; \psi)$ [193].

THEOREM 5.1. *Suppose that $\psi\colon \mathbb{R} \to \mathbb{R}^+$ is decreasing with $x\psi(x) \leqslant 1/2$ for large x and let $s_0 = \sup\{\beta\colon \lim_{x\to\infty} x\psi(x)^s = \infty\}$. If (R, ν) is a regular system then*

$$\dim \Lambda(R, \nu; \psi) \geqslant s_0.$$

As an application of the theorem, we will obtain the correct lower bound of the Hausdorff dimension of the set

$$\mathcal{K}_v = \{\xi\colon |\xi - p/q| < q^{-v-1} \text{ for infinitely many } p/q \in \mathbb{Q}\}$$

of v-approximable numbers (see §1.3). Let $r = p/q$ where p, q are coprime and $\nu(p/q) = q^2$, so that $(\mathbb{Q}, \nu)$ is a regular system. Choose $\psi(x) = x^{-(v+1)/2}$ and $0 < s < 2/(v+1) < 1$. Then

$$x\psi(x)^s = x^{1-(v+1)\,s/2}$$

and so $s_0 = 2/(v+1)$. It follows that when $v > 1$, $\dim \mathcal{K}_v \geqslant 2/(v+1)$; this downward inequality is the hard part of the Jarník-Besicovitch theorem which now follows from (3.10). Some other applications and refinements of previous work, including extensions to simultaneous Diophantine approximation in the plane, are now given.

5.2.1. Regular systems of algebraic numbers. Recall that $\mathbb{A}_n$ is the set of real algebraic numbers α of degree at most n and let $n \geqslant 2$. In their study of $\mathfrak{K}_w$ (see [13] and Chapter 4, (4.1)), Baker and Schmidt showed that $(\mathbb{A}_n, \nu)$, where $\nu(\alpha) = h(\alpha)^{n+1} (\log h(\alpha))^{-3n(n+1)}$, is a regular system. In fact $(\mathbb{A}_n, \nu)$ is a regular system when

$$\nu(\alpha) = \left(\frac{h(\alpha)}{\log h(\alpha)}\right)^{n+1}.$$

To see this, let I be an interval. For each positive integer n, let $A_I(N) = A_I^{(n)}(N)$ denote the set of all $\xi \in I$ for which there exists a real algebraic number α of degree at most n and height $h(\alpha) \leqslant N$ such that

$$|\xi - \alpha| < N^{-n-1} (\log N)^{n+1}.$$

We begin by proving that the set $A_I(N)$ approximates I in measure.

LEMMA 5.2. *$|A_I(N)| \to |I|$ as $N \to \infty$.*

PROOF. Let $B_I(N)$ be the set of $\xi \in I$ for which there exists an integral polynomial P of degree at most n and height at most N such that

$$0 < |P(\xi)| < N^{-n+1}\psi(N),$$

where $\psi(q), q \in \mathbb{N}$, is a positive monotonic decreasing sequence with $\sum_q \psi(q)$ convergent. By Theorem 2.1, $|\bigcap_{N=1}^{\infty} \bigcup_{M=N}^{\infty} B_I(M)| = 0$ and hence

$$\left|\bigcup_{M=N}^{\infty} B_I(M)\right| \to 0$$

as $N \to \infty$. Now let ξ be any transcendental number in $I \backslash A_I(N)$. By Minkowski's linear forms theorem [59], there is an integral polynomial P of degree at most n such that

$$0 < |P(\xi)| \leqslant (\sup\{|x| \colon x \in I\} + 1) N^{-n} (\log N)^{(n-1)(1+1/n)}, \tag{5.1}$$

$$|a_1| \leqslant (\sup\{|x| \colon x \in I\} + 1)^{-1} N, \tag{5.2}$$

$$|a_j| \leqslant N(\log N)^{-(1+1/n)}, \quad 2 \leqslant j \leqslant n. \tag{5.3}$$

The height $h(P)$ of P clearly satisfies $h(P) \ll N$ and for N sufficiently large, we can ensure that $h(P) \leqslant N$. The cases of $h(P)$ small and large are considered separately.

(a) Assume that $h(P) < \log N$. Since the root α of P nearest to ξ satisfies (by Lemma 2.8 with $j = n$)

$$|\xi - \alpha| \ll (|P(\xi)|/h(P))^{1/n} \ll N^{-1} \log N$$

and since the number of polynomials P with $h(P) < \log N$ is $\ll (\log N)^{n+1}$, we see that ξ lies in a set of measure $\ll (\log N)^{n+2}/N$, which tends to 0 as $N \to \infty$.

(b) Next assume that $h(P) \geqslant \log N$ and $|P'(\xi)| \ll N(\log N)^{-(1+1/n)}$. Then by (5.1) and (5.3), $h(P) \ll N(\log N)^{-(1+1/n)}$, whence by substituting for N^{-n} in (5.1) and using $h(P) \ll N$,

$$0 < |P(\xi)| \ll h(P)^{-n}(\log h(P))^{-(1+1/n)}.$$

Thus provided that N is sufficiently large, ξ lies in $B_I(h(P))$. It follows that $\xi \in \bigcup_{r\geqslant \log N} B_I(r)$, a set with measure which tends to 0 as $N \to \infty$.

(c) Finally assume that $|P'(\xi)| \gg N(\log N)^{-(1+1/n)}$ (note that $|P'(\xi)| \ll N$). Since $|P''(x)| \ll N$ for $|x| \ll 1$, we have that $|P'(x)| \gg N(\log N)^{-(1+1/n)}$ for each x with $|x-\xi| \ll N^{-1/2}$ say. Using the estimate (5.1) for $|P(\xi)|$ and the mean value theorem, it can be seen by contradiction that $P(x)$ changes sign in the interval $\{x\colon |x-\xi| \ll N^{-1/2}\}$ which thus contains a real root α say. Hence by the mean value theorem,

$$|\xi - \alpha| \ll |P(\xi)|\, N^{-1}(\log N)^{1+1/n} \ll N^{-n-1}(\log N)^{n+1}.$$

Thus if N is large, then $\xi \in A_I(N)$, contradicting the initial assumption that $\xi \notin A_I(N)$. □

It now follows that when $\nu(\alpha) = (h(\alpha)/\log h(\alpha))^{n+1}$, $(\mathbb{A}_n, \nu)$ is a regular system in $\mathbb{R}$. Indeed let I be any finite interval in $\mathbb{R}$. By Lemma 5.2, $|A_I(N)| \geqslant |I|/2$ for all sufficiently large N. Let $\{\gamma_1, \dots, \gamma_t\}$ be a maximal subset of elements of $\mathbb{A}_n$ with $h(\gamma_j) \leqslant N$ and $|\gamma_j - \gamma_k| \geqslant 1/K$ for distinct j, k, where $K = (N/\log N)^{n+1}$. Then each $\gamma \in \mathbb{A}_n$ with $h(\gamma) \leqslant N$ is at most a distance $1/K$ from some γ_j. Thus the intervals $(\gamma_1 - 2/K, \gamma_1 + 2/K), \dots, (\gamma_t - 2/K, \gamma_t + 2/K)$, which have measure at most $4t/K$, cover $A_I(N)$, whence $|A_I(N)| \leqslant 4t/K$ and $t \geqslant |I|\, K/8$.

The lower bound for the Hausdorff dimension of the set $\mathfrak{K}_w$ is established in two steps. Since $(\mathbb{A}_n, \nu)$ is a regular system by the preceding lemma, Theorem 5.1 above with $\psi(x) = x^{-w/(n+1)} \log x$ gives

$$\dim \Lambda(\mathbb{A}_n, \nu; \psi) \geqslant (n+1)/w.$$

Hence the set of $\xi \in \mathbb{R}$ for which there exist infinitely many real algebraic numbers α with degree at most n such that $|\xi - \alpha| < \psi(\nu(\alpha))$ has Hausdorff dimension at least $(n+1)/w$. But by the definition of ψ and ν,

$$\psi(\nu(\alpha)) = \nu(\alpha)^{-w/(n+1)} \log \nu(\alpha) \ll h(\alpha)^{-w}(\log h(\alpha))^{w+1}$$

and so $\xi \in \mathfrak{K}_w$. Thus $\dim \mathfrak{K}_w \geqslant (n+1)/w$ and since $\mathfrak{K}_{v+1} \subset \mathscr{M}_v$, it follows that $\dim \mathscr{M}_v \geqslant (n+1)/(v+1)$ for $v > 1$ [13]. Note that the case $n = 1$ corresponds to the Jarník-Besicovitch theorem. The choice of the function g permits the deduction that $\dim \mathfrak{K}'_w = \dim \mathfrak{K}_w$ (see §3.5.6).

We now consider an extension of the above to the simultaneous case with different error functions, discussed in [43, Chapter 2, §4]. First the notion of regular systems is extended to the plane. The triple (R, μ, ν), where R is a countable set of points in $\mathbb{R}^2$ and $\mu, \nu\colon R \to \mathbb{R}^+$ are functions, is called a *regular system of points in the plane* if there exist constants $c_1 = c_1(R, \mu, \nu)$ and $c_2 = c_2(R, \mu, \nu)$ such that for any rectangle $I \times J$, there exist positive constants $K(I \times J), L(I \times J)$ such that for any integers $K \geqslant K(I \times J)$, $L \geqslant L(I \times J)$ with $(\log K)/(\log L) = c_2$, there exist $(r_1, s_1), \ldots, (r_t, s_t)$ in $R \cap (I \times J)$ such that for any distinct i, j, $1 \leqslant i, j \leqslant t$, and (r_i, s_i), (r_j, s_j), we have

$$\left.\begin{array}{l} \mu(r_i) \leqslant K,\ \nu(s_i) \leqslant L,\ 1 \leqslant i \leqslant t, \\ |r_i - r_j| \geqslant K^{-1} \text{ or } |s_i - s_j| \geqslant L^{-1}, \\ t \geqslant c_1 KL|I \times J|. \end{array}\right\} \tag{5.4}$$

Higher dimensional regular systems are defined analogously. The closely related well distributed systems were introduced by M. Melián and D. Pestana to study analogues of the Jarník-Besicovitch theorem in hyperbolic space (see §5.2 in the Notes and [158]). I. R. Dombrowsky [97] has used planar regular systems to obtain the correct lower bound for the Hausdorff dimension of the set $\mathfrak{K}(w_1, w_2)$; the upper bound was obtained in §4.1.2.

Let P denote an integral irreducible polynomial with deg $P \leqslant n$ and with height $h(P)$. Recall that the set $\mathfrak{K}(w_1, w_2)$ consists of points $(\xi_1,\ \xi_2) \in (0,1)^2$ satisfying the inequalities

$$|\xi_1 - \alpha_1| < h(P)^{-w_1}, \quad |\xi_2 - \alpha_2| < h(P)^{-w_2} \tag{5.5}$$

simultaneously for infinitely many points $(\alpha_1,\ \alpha_2)$, where α_1, α_2 are zeros of some polynomial P.

A lower bound for the Hausdorff dimension of $\mathfrak{K}(w_1, w_2)$ is obtained by constructing a planar regular system consisting of the set R say of real algebraic points (α_1, α_2), where α_1, α_2 are real roots of an integral polynomial P of degree at most n, and a suitable pair of functions ν_1, ν_2. This bound coincides with the upper bound obtained in §4.1.2, giving the Hausdorff dimension. Note that the case $w_1 \leqslant w_2$ follows by an obvious change in the statement. The construction is similar to the proof of Baker and Schmidt [13] and we state the two main lemmas only. The first is another measure result.

LEMMA 5.3. *Let G be a rectangle in the plane and let $R(N)$ be the set of points (α_1, α_2) in G where α_1, α_2 are real roots of an integral polynomial P of degree at most n and with $h(P) \leqslant N$. For given $\varepsilon > 0$, let β_1, β_2 satisfy*

$$\min\{\beta_1, \beta_2\} > 1 + \varepsilon,\ \beta_1 + \beta_2 = n + 1.$$

Denote by $B(N)$ the set of those vectors $(\omega_1, \omega_2) \in G$ for which there exists (α_1, α_2) in $R(N)$ with

$$|\omega_1 - \alpha_1| < N^{-\beta_1+\varepsilon}, \quad |\omega_2 - \alpha_2| < N^{-\beta_2+\varepsilon}.$$

Then $|B(N)| \to |G|$ as $N \to \infty$.

We turn to establishing the lower bound for dim $\mathfrak{K}(w_1, w_2)$. It can be shown that $(R; \nu_1, \nu_2)$, where $\nu_i(\alpha_i) = h(\alpha_i)^{\beta_i - \varepsilon}$, $i = 1, 2$, is a regular system in the plane. Let $\psi_1, \psi_2 \colon \mathbb{R}^+ \to \mathbb{R}^+$ be a pair of positive functions. Denote by $\Lambda(R; \nu_1, \nu_2; \psi_1, \psi_2)$

the set of those real points (ξ_1, ξ_2) for which there are infinitely many $(\alpha_1, \alpha_2) \in R$ such that

$$|\xi_1 - \alpha_1| < \psi_1(\nu_1(\alpha_1)), \quad |\xi_2 - \alpha_2| < \psi_2(\nu_2(\alpha_2)).$$

The following lemma was proved by Dombrovsky [97].

LEMMA 5.4. *Let $\sigma_1 > \sigma_2 \geqslant 1$ and let $\psi_i(x) = x^{-\sigma_i}$, $i = 1, 2$. Then*

$$\dim \Lambda(R; \nu_1, \nu_2; \psi_1, \psi_2) \geqslant \begin{cases} (2 + \sigma_1 - \sigma_2)/\sigma_1, 1 \leqslant \sigma_2 < 2 \\ 2/\sigma_2, \qquad \sigma_2 \geqslant 2. \end{cases}$$

Now we put $\beta_1 = \beta_2 = (n+1)/2, \sigma_1 = w_1/(\beta_1 - \varepsilon), \sigma_2 = w_2/(\beta_2 - \varepsilon)$. Then the inclusion

$$\Lambda(R, \nu_1, \nu_2; \psi_1, \psi_2) \subset \mathfrak{K}(w_1, w_2),$$

holds. Hence

$$\dim \mathfrak{K}(w_1, w_2) \geqslant \begin{cases} (n + 1 + w_1 - w_2 - 2\varepsilon)/w_1, & (n+1)/2 \leqslant w_2 < n+1, \\ (n + 1 - 2\varepsilon)/w_2, & n + 1 \leqslant w_2. \end{cases}$$

When $1 \leqslant w_2 \leqslant (n+1)/2$, we put $\beta_2 = w_2$, $\beta_1 = n + 1 - w_2$, $\sigma_1 = w_1/(\beta_1 - \varepsilon)$, $\sigma_2 = 1$ and get

$$\dim \mathfrak{K}(w_1, w_2) \geqslant (n + 1 + w_1 - w_2 - 2\varepsilon)/w_1.$$

As $\varepsilon > 0$ is arbitrary, the last two inequalities and Lemma 4.2 give the following.

THEOREM 5.5. *Assume that $w_1 \geqslant w_2 \geqslant 1$ and $w_1 + w_2 > n + 1$. Then*

$$\dim \mathfrak{K}(w_1, w_2) = \begin{cases} (n + 1 + w_1 - w_2)/w_1, & 1 \leqslant w_2 < n+1, \\ (n+1)/w_2, & w_2 \geqslant n+1. \end{cases}$$

5.2.2. Asymptotic formulae and regular systems. Let $a_i, i = 1, 2, \ldots$, be a sequence of numbers in $[0, 1]$. For the arbitrary interval $I \subset [0, 1]$, denote by $\mathcal{N}(Q; I)$ the number of points a_i, $1 \leqslant i \leqslant Q$, lying in I. The sequence (a_i) is said to be *uniformly distributed* (see [59]) if

$$\lim_{Q \to \infty} \frac{\mathcal{N}(Q; I)}{Q} = |I|.$$

We shall now restrict ourselves to sequences (a_i) in $[0, 1]$ for which the asymptotic formula

$$\mathcal{N}(Q; I) = Q\,|I| + O(Q^\lambda), \tag{5.6}$$

where $0 < \lambda < 1$, holds. Suppose $|\mathcal{N}(Q; I) - Q|I|| < c\,Q^\lambda$ for some $c > 0$. Then for each interval I with $|I| > (c+1)Q^{-1+\lambda}$, $\mathcal{N}(Q; I)$ is strictly positive and so I contains at least one element from $a_1, \ldots, a_Q$. We will now construct a regular system from any sequence obeying (5.6).

Let $\nu(a_i) = i^{1-\lambda}$ and let N be sufficiently large. Then the number of elements in the sequence for which $\nu(a_i) \leqslant N$ is $[N^{1/(1-\lambda)}] = Q$ say. Let $K = N/(c+1)$, where c is as above, and let I be an interval. Divide I into subintervals J of equal length

$$|J| = (c+1)Q^{-1+\lambda} > (c+1)N^{-1}.$$

Each such interval contains an element a_i of the sequence. Choose one point from every second subinterval J. Then clearly when $j \neq k$,

$$|a_j - a_k| \geqslant |J| > (c+1)N^{-1} = K^{-1}.$$

The number t of the points a_j on the interval I is at least

$$\frac{|I|}{2|J|} \geqslant \frac{|I|}{2(c+1)} N > \frac{|I|K}{2}.$$

Hence points with an asymptotic distribution in $[0,1]$ with a suitable error term form a regular system. As an application, we consider an example which demonstrates the connection between inhomogeneous approximation and regular systems.

5.2.3. Inhomogeneous approximation and regular systems. When α is irrational, the sequence $(\{q\alpha\} \colon q = 1, 2, \dots)$ is uniformly distributed [59]. When in addition α satisfies the inequality

$$\|q\alpha\| \geqslant \frac{c(\alpha)}{q^w}, \quad w \geqslant 1, \tag{5.7}$$

for all $q \in \mathbb{N}$, the asymptotic formula (5.6) with $c = c(\alpha, \varepsilon)$ and $\lambda = 1 - w^{-1} + \varepsilon$, where $\varepsilon > 0$, holds for the sequence $(\{q\alpha\} \colon q = 1, 2, \dots)$ [147]. Thus the set of points $a_q = \{q\alpha\}$, $q = 1, 2, \dots$, and the function $\nu(a_q) = q^{1-\lambda}$ form a regular system. Hence when $v > 1$, the Hausdorff dimension of the set $V_v(\alpha)$ of points $\beta \in \mathbb{R}$ for which $\|q\alpha - \beta\| < q^{-v}$ holds for infinitely many $q \in \mathbb{N}$ satisfies

$$\frac{1}{wv} \leqslant \dim V_v(\alpha) \leqslant \frac{1}{v}$$

(the upper bound is straightforward). Thus if α satisfies (5.7) with $w = 1 + \varepsilon$ for any $\varepsilon > 0$, then for $v > 1$, $\dim V_v = 1/v$.

Let us note one other connection between regular systems and uniform distribution. Since the points in a regular system are dense, the points of any regular system $\gamma_1, \gamma_2, \dots$, ordered with respect to the value of $\nu(\gamma_i)$, can be shown readily to be a uniformly distributed sequence.

5.2.4. Planar curves. The notation in this sub-section is as in §4.2.2; thus Γ is a C^2 planar curve with non-zero curvature everywhere except on a set of Hausdorff dimension zero. R. C. Baker [16] R. C. Baker [16] used regular systems to show that for $v \geqslant 2$, the Hausdorff dimension of the set $X(\Gamma, v)$ (defined in §4.2.2) satisfies

$$\dim X(\Gamma, v) \geqslant 3/(v+1)$$

and it follows from (4.5) that $\dim X(\Gamma, v) = 3/(v+1)$ and by (1.15) that

$$\dim \mathscr{L}_v(\Gamma) = \dim \mathcal{L}_v(\Gamma) = \dim \mathcal{L}'_v(\Gamma) \geqslant 3/(v+1).$$

This lower bound is covered by Theorem 5.11 below. D. V. Vasilyev has used regular systems to obtain the same lower bound when $v \geqslant 1/2$ for complex curves [221].

5.3. Ubiquitous systems

In order to obtain a lower bound for the Hausdorff dimension of the more general sets occuring in small denominator problems (see Chapter 7), ubiquitous systems were introduced in [90]. The notion was abstracted from [54] which used geometrical ideas based on those in Besicovitch's paper [48] and a mean and variance argument from [59, Chapter 7]. In the one dimensional case and when the resonant sets consist of points, ubiquitous and regular systems are virtually equivalent and essentially differ only in their formulation [193]. Resonant sets (see §1.5.1) play a fundamental role in ubiquity which in essence ensures that they are in good supply. They can be thought of as generalisations of rational numbers and are called resonant because of the connection between rational dependence and the physical phenomenon of resonance (further details are in Chapter 7). They are of a relatively simple nature, being finite unions of points or parts of lines, curves, planes, surfaces and so on, which are solution sets of Diophantine equations. The sets, such as $S(M;\psi)$ and $L(M;\psi)$, are lim-sup sets of sequences of neighbourhoods of resonant sets in the parameter space U.

With a view to application to manifolds, we take U to be a non-empty open subset of $\mathbb{R}^m$. Let

$$\mathscr{R} = \{R_j \subset U : j \in J\} \tag{5.8}$$

be a family of sets, which we call 'resonant', indexed by J, where each $j \in J$ has a weight $\lfloor j \rfloor > 0$. Let the function $\rho\colon \mathbb{N} \to \mathbb{R}^+$ converge to 0 as $N \to \infty$ and let $A(N)$, $N = 1, 2, \ldots$, be a sequence of subsets of U such that

$$\lim_{N\to\infty} |U \setminus A(N)| = 0. \tag{5.9}$$

Let

$$B(R_j;\delta) = \{u \in U\colon \operatorname{dist}_\infty(u, R_j) < \delta\},$$

where $\operatorname{dist}_\infty(u,R) = \inf\{|u-r|\colon r \in R\}$, the distance from u to R in the supremum norm. Suppose that there exists a constant $d \in [0,n]$ such that given any hypercube $H \subset U$ with $\ell(H) = \rho(N)$ and such that $H/2$ intersects $A(N)$, then there exists a $j \in J$ with $\lfloor j \rfloor \leqslant N$ such that for all $\delta \in (0, \rho(N)]$,

$$|H \cap B(R_j;\delta)| \gg \delta^{m-d}\,\ell(H)^d. \tag{5.10}$$

Suppose further that for any other hypercube H' in U with $\ell(H') \leqslant \rho(N)$,

$$|H' \cap H \cap B(R_j;\delta)| \ll \delta^{m-d}\,\ell(H')^d. \tag{5.11}$$

Then the pair $(\mathscr{R}, \lfloor\cdot\rfloor)$ is called a *ubiquitous system with respect to* ρ (reference to the weight is usually omitted).

The intersection estimates (5.10) and (5.11) have been used in preference to more geometrical descriptions of the intersections $H \cap R_j$ for generality. The first requires that the hypercube H and the resonant set R_j intersect substantially and that small hypercubes H' intersect $H \cap R_j$ as they 'should'. For resonant sets R_j with a reasonable structure, d will be the topological dimension of each R_j and the intersection conditions (5.10) and (5.11) will be satisfied more or less automatically. Indeed when the R_j are d-dimensional affine spaces in Euclidean space, we can take the approximating set $A(N)$ to be a union of $\rho(N)$-neighbourhoods of R_j, namely

$$A(N) = \bigcup_{\lfloor j \rfloor \leqslant N} B(R_j; \rho(N)). \tag{5.12}$$

It is then readily verified that the intersection conditions (5.10) and (5.11) can be replaced by the single measure condition:

$$\left| \bigcup_{\lfloor j \rfloor \leqslant N} B(R_j; \rho(N)) \right| \to |U| \text{ as } N \to \infty. \tag{5.13}$$

Ubiquity can be relatively simple to establish and in practice the function ρ emerges naturally. For instance Dirichlet's theorem implies that the set of rationals (with weight the modulus of the denominator) is ubiquitous with respect to a function comparable with $N^{-2} \log N$; in higher dimensions the rational points $\mathbf{p}/q$, $\mathbf{p} \in \mathbb{Z}^n$, $q \in \mathbb{N}$, are ubiquitous with respect to a function asymptotic to $N^{-1-1/n} \log N$ [90].

For linear forms, the resonant sets $R_{\mathbf{q}}$,where $\mathbf{q} \in \mathbb{Z}^m \setminus \{\mathbf{0}\}$, given by

$$R_{\mathbf{q}} = \{A \in [0,1]^{mn} \colon \mathbf{q}A = \mathbf{r} \text{ for some } \mathbf{r} \in \mathbb{Z}^n\}$$

are ubiquitous with respect to the function $mN^{-1-m/n} \log N$ and 'most' matrices $A \in [0,1]^{mn}$ are 'close' to a set $R_{\mathbf{q}}$ with weight $\lfloor \mathbf{q} \rfloor = |\mathbf{q}|$ not too large [82]. To see this, consider the δ-neighbourhood of $R_{\mathbf{q}}$, where $|\mathbf{q}|, \delta > 0$, given by

$$B(R_{\mathbf{q}}; \delta) = \{A \in I^{mn} \colon \operatorname{dist}_\infty(A, R_{\mathbf{q}}) < \delta\}.$$

This set is closely related to the neighbourhood $B_\delta(R_{\mathbf{q}}) = \{A \in I^{mn} \colon |\langle \mathbf{q}A \rangle| < \delta\}$. Indeed since $\mathbf{q} \cdot (x - y) = x \cdot \mathbf{q} - \mathbf{r}$ when $\mathbf{q} \cdot y = \mathbf{r}$,

$$B_{\delta|\mathbf{q}|}(R_{\mathbf{q}}) \subseteq B(R_{\mathbf{q}}; \delta) \subseteq B_{\delta m|\mathbf{q}|}(R_{\mathbf{q}}) \tag{5.14}$$

for each non-zero $\mathbf{q}$. By (5.13), it suffices to show that

$$\lim_{N \to \infty} \left| \bigcup_{1 \leqslant |\mathbf{q}| \leqslant N} B(R_{\mathbf{q}}; \rho(N)) \right| = 1, \tag{5.15}$$

where $\rho(N) = m^{-1} N^{-1-m/n} \log N$. It follows from Dirichlet's theorem [201] that

$$[0,1]^{mn} = \bigcup_{1 \leqslant |\mathbf{q}| \leqslant N} B_{N^{-m/n}}(R_{\mathbf{q}}), \tag{5.16}$$

since for each A in $[0,1]^{mn}$ and each $N \geqslant 2$, there exist integer vectors $\mathbf{q} \in \mathbb{Z}^m$ with $1 \leqslant |\mathbf{q}| \leqslant N$ and $\mathbf{r} \in \mathbb{Z}^n$ such that $|\mathbf{q}A - \mathbf{r}| < N^{-m/n}$. Now

$$\bigcup_{1\leqslant|\mathbf{q}|\leqslant N} B(R_{\mathbf{q}};\rho(N)) = S(N) \cup \bigcup_{N/\log N\leqslant|\mathbf{q}|\leqslant N} B(R_{\mathbf{q}};\rho(N))$$

where $S(N) = \bigcup_{1\leqslant|\mathbf{q}|<N/\log N} B(R_{\mathbf{q}};\rho(N))$ is the set of 'small denominators'. But by (5.14) and the choice of ρ,

$$S(N) \subseteq \bigcup_{1\leqslant|\mathbf{q}|<N/\log N} B_{\rho(N)m|\mathbf{q}|}(R_{\mathbf{q}}) = \bigcup_{1\leqslant|\mathbf{q}|<N/\log N} B_{N^{-m/n}}(R_{\mathbf{q}}).$$

Using the periodicity of $\langle\mathbf{q}A\rangle$, it can be shown that $|B_\delta(R_{\mathbf{q}})| = 2^n\delta^n$ (see [59], [96] or [210]). Hence

$$|S(N)| \leqslant \sum_{1\leqslant r<N/\log N} \sum_{|\mathbf{q}|=r} (2N^{-m/n})^n \ll N^{-m} \sum_{1\leqslant r<N/\log N} r^{m-1} \ll (\log N)^{-m}.$$

Evidently the complement of $S(N)$ in $[0,1]^n$ contains

$$\bigcup_{N/\log N\leqslant|\mathbf{q}|\leqslant N} B_{N^{-m/n}}(R_{\mathbf{q}})$$

and so by (5.16) has measure $1 + O(\log N)^{-m}$. Thus (5.15) holds and the family $\{R_{\mathbf{q}}\colon \mathbf{q} \in \mathbb{Z}^m \setminus \{0\}\}$ is ubiquitous with respect to $\rho(N) = m^{-1}N^{-1-m/n}\log N$. For another approach see [83].

5.3.1. A general lower bound. The distribution of the resonant sets in ubiquitous systems allows the determination of a general lower bound for the lim-sup set

$$\Lambda(\mathscr{R}\,;\psi) = \{u \in U\colon\ \mathrm{dist}_\infty(u, R_j) < \psi(\lfloor j\rfloor) \text{ for infinitely many } j \in J\},$$

where $\psi\colon \mathbb{N} \to \mathbb{R}^+$ is a decreasing function and the resonant sets have common dimension $d = \dim\mathscr{R}$ say and codimension $m - d = \operatorname{codim}\mathscr{R}$.

THEOREM 5.6. *Suppose $\mathscr{R}$ is a family of resonant sets which is ubiquitous with respect to ρ and that $\psi\colon \mathbb{R}^+ \to \mathbb{R}^+$ is a decreasing function satisfying $\psi(N) \leqslant \rho(N)$ for N sufficiently large. Then*

$$\dim\Lambda(\mathscr{R}\,;\psi) \geqslant \dim\mathscr{R} + \gamma\operatorname{codim}\mathscr{R},$$

where $\gamma = \limsup_{N\to\infty} \log\rho(N)/\log\psi(N) \leqslant 1$.

This is proved in [90] but another proof is given using the mass distribution principle (see §3.5.7); some details in common will be omitted. The constant $\gamma \leqslant 1$ since $\psi(N) \leqslant \rho(N)$ for N sufficiently large.

The hypercube U is divided up into congruent hypercubes with disjoint interiors. For each positive integer N, consider the collection of hypercubes $H \subset U$ with $\ell(H) = \rho(N)$ and vertices lying on the lattice $\rho(N)\mathbb{Z}^m$. Let $\mathcal{G}(U)$ be the collection of *good* hypercubes H for which $H/4$ (λH is H shrunk by λ) meets the resonant set

R_j for some j, $\lfloor j \rfloor \leqslant N$ (thus H meets $A(N)$ 'substantially'). It follows from (5.9) that

$$\#\mathcal{G}(U) \asymp \rho(N)^{-m}|U|. \tag{5.17}$$

For each hypercube H in $\mathcal{G}(U)$, there exists a $j \in J$, $\lfloor j \rfloor \leqslant N$, satisfying the ubiquity intersection conditions (5.10) and (5.11).

Consider the set of *deleted hypercubes* D in $H/4$, of sidelength $\ell(D) = \psi(N)/2$ and volume $|D| \asymp \psi(N)^m$, centred on the resonant set R_j with centres of the nearest neighbours a distance $2^m\psi(N)$ apart, at least $2\psi(N)$ from the boundary of $R \cap (H/4)$; and with R_j deleted (this is done to guarantee infinitely many different j). Since ψ is decreasing, each point u in a deleted hypercube D satisfies

$$0 < \operatorname{dist}_\infty(u, R_j) \leqslant \psi(N)/2 < \psi(\lfloor j \rfloor). \tag{5.18}$$

The number of such deleted hypercubes is comparable to $(\rho(N)/\psi(N))^d$. Let

$$T = T(N) = \bigcup_{H \in \mathcal{G}(U)} \bigcup_{D \subseteq H/4} D$$

be the union of all these deleted hypercubes D lying in $H/4$ for some good hypercube H in U. Their number t say is given by

$$t = \sum_{H \in \mathcal{G}(U)} \sum_{D} 1 \asymp \left(\frac{\rho(N)}{\psi(N)}\right)^d |U|\, \rho(N)^{-m}.$$

The following simple counting result is Lemma 1 in [90].

LEMMA 5.7. *For any set X in U with null boundary, there exists an integer $N^*(X)$ such that for all $N \geqslant N^*(X)$,*

$$|T \cap X| \asymp t\,|X|.$$

We now start a Cantor-type construction using a repeated subdivision idea which goes back to Jarník and Besicovitch. Choose N_1 sufficiently large to ensure that Lemma 5.7 holds. Then

$$T_1 = \bigcup_{H \in \mathcal{G}_1(U)} \bigcup_{D} D, \quad t_1 \asymp \rho(N_1)^{d-m}\, \psi(N_1)^{-d}, \tag{5.19}$$

where $|U|$ is absorbed into the implied constant and D is a *level 1* deleted hypercube of sidelength $\psi(N_1)/2$. This completes the first level of the Cantor construction.

For the second level, the cube U is further sub-divided into hypercubes H with $\ell(H) = \rho(N_2)$ and vertices lying on the lattice $\rho(N_2)\mathbb{Z}^m$; we are only interested in the good hypercubes which lie in T_1. Let $\mathcal{G}_2$ be the collection of good hypercubes for which $H/4$ meets $A(N_2)$. Choose $N_2 > N_1$ large enough to ensure that Lemma 5.7 holds for each level 1 deleted hypercube D in T_1. Then for such D,

$$\#\mathcal{G}_2(D) \gg \rho(N_2)^{-m}\,|D|. \tag{5.20}$$

Let T_2 be the set of level 2 deleted hypercubes D' ($\ell(D') = \psi(N_2)/2$) in $H/4$ where $H \in \mathcal{G}_2(T_1)$ and centred on a resonant set R with centres of the nearest neighbours a distance $2^m\psi(N_2)$ apart and at least $2\psi(N_2)$ from the boundary of $R \cap (H/4)$. From (5.20) and since $\ell(D) = \psi(N_1)/2$,

$$t_2 \asymp \rho(N_2)^{-m}\left(\frac{\rho(N_2)}{\psi(N_2)}\right)^d t_1|D| \asymp \frac{\rho(N_2)^{d-m}\psi(N_1)^m}{\psi(N_2)^d}\, t_1. \tag{5.21}$$

Choose any subsequence (q_j) so that

$$\lim_{j\to\infty}\frac{\log\rho(q_j)}{\log\psi(q_j)} = \gamma.$$

The argument now proceeds by constructing the r-th level of the Cantor set from the $(r-1)$-th level. Choose $N_r = q_j$ for j sufficiently large to ensure that Lemma 5.7 and the following hold.

(a) The number $\#\mathcal{G}_r(D)$ of 'good' cubes of sidelength $\rho(N_r)$ contained in each deleted cube $D \in T_{r-1}$ satisfies

$$\#\mathcal{G}_r(D) \gg \psi(N_{r-1})^m\rho(N_r)^{-m}, \tag{5.22}$$

(b)

$$\psi(N_r) \leqslant \prod_{i=1}^{r-1}(\psi(N_i)/\rho(N_i))^{r(m-d)}. \tag{5.23}$$

Now define T_r recursively as follows:

$$T_r = \bigcup_{H\in\mathcal{G}_r(T_{r-1})}\ \bigcup_{D\subset H/4} D \subset T_{r-1}$$

for each $r = 2, 3, \ldots$ as above. Then it can be verified that

$$\begin{aligned} t_r &= \#\mathcal{G}_r(T_{r-1})\,\#\{D\colon D\subset H/4, H\in\mathcal{G}_r(T_{r-1})\} \\ &\gg \rho(N_r)^{-m}\left(\frac{\rho(N_r)}{\psi(N_r)}\right)^d t_{r-1}\,\psi(N_{r-1})^m. \end{aligned} \tag{5.24}$$

Since $|D'| \asymp \psi(N_r)^m$ for each $D' \subset T_r$, it follows that

$$\begin{aligned} t_r &\gg \rho(N_r)^{d-m}\rho(N_{r-1})^{d-m}\psi(N_r)^{-d}\psi(N_{r-1})^{m-d}t_{r-2}\,\psi(N_{r-2})^m \\ &\gg \prod_{i=2}^{r}(\rho(N_i)/\psi(N_i))^{d-m}t_1\psi(N_1)^m\psi(N_r)^{-m} \\ &\gg \prod_{i=1}^{r}(\rho(N_i)/\psi(N_i))^{d-m}\psi(N_r)^{-m}, \end{aligned}$$

since $t_1 \gg \rho(N_1)^{d-m}\psi(N_1)^{-d}$. Let

$$T_\infty = \bigcap_{r=1}^{\infty} T_r.$$

Given $u \in T_\infty$, for each $r = 1, 2, \ldots$, there exists j with $\lfloor j \rfloor \leqslant N_r$ such that $0 < \operatorname{dist}\infty(u, R_j) < \psi(N_r)$. But the cubes are deleted and $\psi(N) \to 0$ as $N \to \infty$, so that each $u \in T_\infty$ satisfies (5.18) for infinitely many $j \in J$. Hence

$$T_\infty \subset \Lambda(\mathscr{R}; \psi). \tag{5.25}$$

A probability measure μ supported on the Cantor type subset T_∞ is now constructed. First for each $D \subset T_1$, define

$$\mu_1(D) = 1/t_1\,,$$

so that $\mu_1(U) = 1$. For each $D \subset T_r$, $r \geqslant 2$, define

$$\mu_r(D) = 1/t_r\,,$$

so that $\mu_r(U) = 1$. Note that for each $r = 2, 3, \ldots$, and $D \subset T_{r-1}$,

$$\mu_{r-1}(D) = \sum_{D' \subset T_r \cap D} \mu_r(D') = \frac{1}{t_r}\,\#\{D' \colon D' \subset D\}.$$

On taking the limit as $r \to \infty$, the iterative construction gives rise to a probability measure μ on U and supported on the 'Cantor' set T_∞, so that $\mu(T_\infty) = 1$ (see [100] for details).

To use the mass distribution principle to show that $\dim T_\infty \geqslant s$, it suffices to prove that for each hypercube C in U, $\mu(C) \ll \ell(C)^s$. We consider the measure $\mu_r(D)$ of the deleted hypercube D in T_r. By definition

$$\begin{aligned}\mu_r(D) = 1/t_r &\ll \prod_{i=1}^{r} \rho(N_i)^{m-d}\psi(N_i)^{d-m}\psi(N_r)^m \\ &\ll \rho(N_r)^{m-d}\psi(N_r)^d \prod_{i=1}^{r-1} (\rho(N_i)/\psi(N_i))^{m-d} \\ &\ll \rho(N_r)^{m-d}\psi(N_r)^{d-\frac{1}{r}},\end{aligned}$$

by the choice of N_r and (5.23). Hence $\mu_r(D) \ll \psi(N_r)^{s_r}$, where

$$s_r = \frac{(m-d)\log\rho(N_r) + (d - 1/r)\log\psi(N_r)}{\log\psi(N_r)} = d - \frac{1}{r} + (m-d)\,\gamma_r$$

and $\gamma_r = (\log\rho(N_r))/(\log\psi(N_r))$, so that $\gamma = \lim_{r\to\infty}\gamma_r$. Choose r sufficiently large so that $s_r > s$, where for the rest of the section

$$s = d + (m-d)\gamma - \eta.$$

It follows that for each D in T_r, the measure μ satisfies

$$\mu(D) = 1/t_r \ll \psi(N_r)^s. \tag{5.26}$$

LEMMA 5.8. *For any sufficiently small hypercube C in U, $\mu(C) \ll \ell(C)^s$.*

PROOF. If r is sufficiently large, then by Lemma 5.7, we can count deleted hypercubes D in $T_r \cap C$ to get that

$$\mu(C) = \mu(T_\infty \cap C) \ll \left(\frac{\rho(N_r)}{\psi(N_r)}\right)^d \left(\frac{\ell(C)}{\rho(N_r)}\right)^m \mu(D)$$

for $\rho(N_r) < \ell(C)$ and

$$\mu(C) \ll \left(\frac{\ell(C)}{\psi(N_r)}\right)^d \mu(D)$$

otherwise. Let C be small enough so that the r for which $\psi(N_r) < \ell(C) \leqslant \psi(N_{r-1})$ is sufficiently large. By (5.24) and (5.26) applied to $\mu(D)$ and to $1/t_{r-1}$, when $\rho(N_r) < \ell(C)$,

$$\begin{aligned}\mu(C) &\ll \frac{t_r}{t_{r-1}} \left(\frac{\ell(C)}{\psi(N_{r-1})}\right)^m \mu(D) \ll \frac{1}{t_{r-1}} \left(\frac{\ell(C)}{\psi(N_{r-1})}\right)^m \\ &\ll \ell(C)^m \psi(N_{r-1})^{s-m} \ll \ell(C)^s,\end{aligned}$$

since $s \leqslant m$. If $\rho(N_r) \geqslant \ell(C)$, then by (5.24), $\mu(C) \ll \ell(C)^d \psi(N_r)^{s-d} \ll \ell(C)^s$. □

It follows from the mass distribution principle (Lemma 3.12) that

$$\dim T_\infty \geqslant s = d + \gamma(m - d) - \eta.$$

But $\eta > 0$ is arbitrary, whence $\dim T_\infty \geqslant d + \gamma(m - d)$, and Theorem 5.6 follows from (5.25).

As an application, we shall prove the complementary result to Lemma 4.1.

LEMMA 5.9. *Let $\psi: \mathbb{N} \to \mathbb{R}^+$ be a decreasing function and let λ be the lower order of $1/\psi$. Then*

$$\dim W(\psi) \geqslant \begin{cases} (m-1)n + (m+n)/(\lambda+1) & \text{when } \lambda \geqslant m/n, \\ mn & \text{when } \lambda \leqslant m/n. \end{cases}$$

Let $\tilde{\psi}(q) = \psi(q)/mq$ for each positive integer q so that the function $\tilde{\psi}: \mathbb{N} \to \mathbb{R}^+$ is decreasing. Let $\tilde{\lambda}$ be the lower order of $1/\tilde{\psi}$ and let

$$\begin{aligned}\Lambda(\tilde{\psi}) &= \{A \in I^{mn} : |A - R_{\mathbf{q}}| < \tilde{\psi}(|\mathbf{q}|) \text{ for infinitely many } \mathbf{q} \in \mathbb{Z}^m\} \\ &= \bigcap_{N=1}^{\infty} \bigcup_{|\mathbf{q}|=N}^{\infty} B(R_{\mathbf{q}}; \tilde{\psi}(|\mathbf{q}|)).\end{aligned}$$

Now $W(\psi)$ can be expressed as the lim-sup set

$$W(\psi) = \bigcap_{N=1}^{\infty} \bigcup_{|\mathbf{q}|=N}^{\infty} B_{\psi(|\mathbf{q}|)}(R_{\mathbf{q}})$$

and so by (5.14), $\Lambda(\tilde{\psi}) \subseteq W(\psi)$. Hence $\dim W(\psi) \geqslant \dim \Lambda(\tilde{\psi})$. By the above, the family $\mathcal{F} = \{R_{\mathbf{q}} \colon \mathbf{q} \in \mathbb{Z}^m \setminus \{\mathbf{0}\}\}$ is ubiquitous with respect to $\rho(N) = mN^{-1-m/n} \log N$. It follows from Theorem 5.6 above that

$$\dim \Lambda(\tilde{\psi}) \geqslant \dim \mathcal{F} + \gamma \operatorname{codim} \mathcal{F},$$

where $\dim \mathcal{F} = (m-1)n$ and $\operatorname{codim} \mathcal{F} = n$ are the common dimension and codimension respectively of the resonant sets $R_{\mathbf{q}}$ in I^{mn}, and where

$$\gamma = \min\left\{1, \limsup_{N\to\infty} \frac{\log \rho(N)}{\log \tilde{\psi}(N)}\right\} = \min\left\{1, \left(1+\frac{m}{n}\right)\frac{1}{\tilde{\lambda}}\right\}.$$

Hence

$$\dim \Lambda(\tilde{\psi}) \geqslant \min\left\{mn, (m-1)n + \left(1+\frac{m}{n}\right)\frac{n}{\tilde{\lambda}}\right\}. \tag{5.27}$$

Now by definition, $\tilde{\psi}(N) = \psi(N)/mN$ so that $\tilde{\lambda} = \lambda + 1$. Substituting in (5.27) gives

$$\dim W(\psi) \geqslant \dim \Lambda(\tilde{\psi}) \geqslant \begin{cases} (m-1)n + (m+n)/(\lambda+1) & \text{when } \lambda \geqslant m/n, \\ mn & \text{when } \lambda \leqslant m/n, \end{cases}$$

which with Lemma 4.1 gives a general form of the Jarník-Besicovitch theorem from which the Hausdorff dimension of the sets $\mathscr{K}_v$, $\mathscr{S}_v(\mathbb{R}^n)$ and $\mathscr{L}_v(\mathbb{R}^n)$ can be deduced readily.

THEOREM 5.10. *Let $\psi\colon \mathbb{N} \to \mathbb{R}^+$ be a decreasing function and let λ be the lower order of $1/\psi$. Then*

$$\dim W(\psi) = \begin{cases} (m-1)n + (m+n)/(\lambda+1) & \textit{when } \lambda \geqslant m/n, \\ mn & \textit{when } \lambda \leqslant m/n. \end{cases}$$

5.3.2. Hausdorff dimension and extremal manifolds. A C^3 planar curve (corresponding to $n = 2$, $m = 1$) with non-vanishing curvature everywhere except on a set with zero Hausdorff dimensionis extremal (by [198]) and

$$\dim \mathscr{L}_v(M) = 3/(v+1)$$

for $v \geqslant 2$ by [16]. In [89] it is shown for manifolds M with dimension $m \geqslant 2$ and 2-curved everywhere except on a set of Hausdorff dimension at most $m-1$ that

$$\dim \mathscr{L}_v(M) = m - 1 + (n+1)/(v+1)$$

for $v \geqslant n$. Ubiquity will now be used to show that the right hand side of the above equation is a general lower bound for the Hausdorff dimension of $\mathscr{L}(M;\psi)$ when M is a C^1 extremal manifold in $\mathbb{R}^n$ (the proof is drawn from [76]). Recall from (1.10) that

$$\mathscr{L}(M\,;\psi) = \{x \in M\colon \|\mathbf{q}\cdot x\| < \psi(|\mathbf{q}|) \text{ for infinitely many } \mathbf{q} \in \mathbb{Z}^n\}.$$

LEMMA 5.11. *Let $\psi\colon \mathbb{N} \to \mathbb{R}^+$ be decreasing with the lower order denoted by λ. Let M be a C^2 extremal manifold embedded in $\mathbb{R}^n$ and suppose $\lambda \geqslant n$. Then*

$$\dim \mathscr{L}(M;\psi) \geqslant m - 1 + (n+1)/(\lambda+1).$$

We choose a Monge domain U to be a hypercube with $\ell(U)$ sufficiently small and with parametrisation $\theta = 1_U \times \varphi$. In view of §3.6, we consider $\dim L(\psi)$. The resonant sets in U are of the form

$$R_{p,\mathbf{q}} = \{u \in U\colon \mathbf{q}\cdot\theta(u) = p\} = \{u \in U\colon \theta(u) \in \Pi_{p,\mathbf{q}}\},$$

where $\mathbf{q} \in \mathbb{Z}^n \setminus \{0\}$ and $p \in \mathbb{Z}$. Using the geometry of numbers we will choose integer vectors $\mathbf{q}$ close to parallel to the plane $\mathbb{R}^m \times \{0\}$, and so not close to being orthogonal to the tangent plane $T_\xi M$ for each $\xi \in M_U$. This ensures that the hyperplanes $\Pi_{p,\mathbf{q}}$ are not close to tangential to M_U.

Let $\eta > 0$ be arbitrary and $N \in \mathbb{N}$ be sufficiently large. Then by Minkowski's linear forms theorem, for each $u \in U$ there exist a vector $\mathbf{q} = \mathbf{q}(u) \in \mathbb{Z}^n$ with $1 \leqslant |\mathbf{q}| \leqslant N$ and an integer $p = p(u)$ such that

$$\left.\begin{aligned} |\mathbf{q}\cdot\theta(u) - p| &\leqslant N^{-n+k\eta}(\log N)^k, \\ |q_i| &\leqslant N,\ i = 1,\dots,m, \\ |q_{m+j}| &\leqslant N^{1-\eta}(\log N)^{-1},\ j = 1,\dots,k \end{aligned}\right\} \tag{5.28}$$

(recall that we can assume without loss of generality that $M_U \subset [-1,1]^n$). In view of Theorem 5.6, it suffices to find a sequence of suitable subsets $A(N)$ which approximate U in Lebesgue measure and a function $\rho\colon \mathbb{N} \to \mathbb{R}^+$ which satisfy the intersection conditions above. By (5.28), the set U can be decomposed for each N as (the vector $\mathbf{q}$ satisfies (5.28))

$$U = A(N) \cup S(N) \cup E(N), \tag{5.29}$$

where

$$E(N) = \{u \in U\colon \operatorname{dist}_\infty(u, \partial U) \leqslant 1/N\}$$

is the set of points close to the boundary of U, where $S(N)$ is the set of points $u \in U$ for which there exists a non-zero solution $p, \mathbf{q}$ of (5.28) with $1 \leqslant |\mathbf{q}| \leqslant N^{1-\eta}$ (and is the set of $u \in U$ with 'denominators' $\mathbf{q}$ which are small compared to the bound in (5.28)) and where

$$A(N) = U \setminus (E(N) \cup S(N)).$$

Thus each $u \in A(N)$ is at least $1/N$ from ∂U (in the supremum metric) and has a 'large' denominator $\mathbf{q}$ satisfying (5.28) with $N^{1-\eta} \leqslant |\mathbf{q}| \leqslant N$. The measure of $E(N)$ converges to 0 as $N \to \infty$ since

$$|E(N)| = |\{u \in U\colon \operatorname{dist}(u, \partial U) \leqslant 1/N\}| \ll \ell(U)^m - (\ell(U) - 1/N)^m \ll N^{-1}.$$

For each $u \in A(N)$, the angle α which $\mathbf{q} = \mathbf{q}(u)$ makes with $\mathbb{R}^m \times \{0\}$ is small since

$$\cos\alpha = \frac{\mathbf{q}}{|\mathbf{q}|_2} \cdot \frac{(q_1, \dots, q_m, 0, \dots, 0)}{(q_1^2 + \dots + q_m^2)^{1/2}} = 1 - O\left(\frac{1}{\log N}\right)^2$$

and hence the hyperplane $\Pi_{p,\mathbf{q}} = \{x \in \mathbb{R}^n : \mathbf{q} \cdot x = p\}$ is not close to being tangential to M_U. Moreover the distance of the boundary of the submanifold

$$\{x \in M : \operatorname{dist}_2(x, \Pi_{p,\mathbf{q}}) < \delta\}$$

from $\Pi_{p,\mathbf{q}}$ is comparable to δ (the constant depends on U and θ).

The set $S(N^{1/(1-\eta)})$ is contained in the set of points $u \in U$ for which there exist a $\mathbf{q} \in \mathbb{Z}^m$ and $p \in \mathbb{Z}$ satisfying

$$|\mathbf{q} \cdot \theta(u) - p| < N^{-(n-k\eta)/(1-\eta)} (\log N)^k (1-\eta)^{-k}$$

with $1 \leqslant |\mathbf{q}| \leqslant N$ and which is a subset of

$$Y(N) = \{u \in U : |\mathbf{q} \cdot \theta(u) - p| < N^{-n-\delta} \text{ for some } \mathbf{q} \in \mathbb{Z}^n, 1 \leqslant |\mathbf{q}| \leqslant N, p \in \mathbb{Z}\},$$

where $0 < \delta < \eta(n-k)/(1-\eta)$ (so that $n + \delta < (n - k\eta)/(1-\eta)$).

LEMMA 5.12. *For any $\delta > 0$,*

$$\limsup_{N\to\infty} Y(N) = \bigcap_{k=1}^{\infty} \bigcup_{N=k}^{\infty} Y(N) \subseteq L_{n+\delta}.$$

PROOF. Recall that

$$L_v = \{u \in U : \|\mathbf{q} \cdot \theta(u)\| < |\mathbf{q}|^{-v} \text{ for infinitely many } \mathbf{q} \in \mathbb{Z}^n\}$$

(see §2.2). Let $u \in \limsup_{N\to\infty} Y(N)$. Then $u \in Y(N_j)$ for an infinite sequence $(N_j : j = 1, 2, \dots)$. Hence for each $j = 1, 2, \dots$, there exist $N_j \in \mathbb{N}$, $\mathbf{q}_j = \mathbf{q}_j(u) \in \mathbb{Z}^n$, $1 \leqslant |\mathbf{q}_j| \leqslant N_j$ and $p_j = p_j(u) \in \mathbb{Z}$ such that

$$|\mathbf{q}_j \cdot \theta(u) - p_j| < N_j^{-n-\delta}.$$

Suppose there are only finitely many different $\mathbf{q}_j$ satisfying the last displayed inequality and let

$$\min\{|\mathbf{q}_j \cdot \theta(u) - p_j| : j \in \mathbb{N}\} = c$$

say. If $c > 0$, then choosing j so that $N_j^{-n-\delta} < c$ gives a contradiction. If $c = 0$, then for each $r \in \mathbb{N}$, $r \leqslant |r\mathbf{q}_j| \leqslant rN_j$ and

$$|(r\mathbf{q}_j) \cdot \theta(u) - (rp_j)| = 0 < (rN_j)^{-n-\delta}.$$

Thus there are infinitely many solutions, contradicting the supposition that there exist only a finite number of different $\mathbf{q}_j$. But $1 \leqslant |\mathbf{q}_j| \leqslant N_j$, whence

$$|\mathbf{q}_j \cdot \theta(u) - p_j| < |\mathbf{q}_j|^{-n-\delta}$$

holds for infinitely many $\mathbf{q}_j$, p_j. Thus $u \in L_{n+\delta}$. □

COROLLARY 5.13. *If M is extremal, then $|S(N)| \to 0$ as $N \to \infty$.*

PROOF. By Fatou's lemma, for any $\delta > 0$

$$\limsup_{N\to\infty} |Y(N)| \leqslant |\limsup_{N\to\infty} Y(N)| \leqslant |L_{n+\delta}| = 0$$

since M is extremal. Thus $\lim_{N\to\infty} |Y(N)| = 0$. But $Y(N) \supseteq S(N^{1/(1-\eta)})$ when $\delta < \eta(n-k)/(1-\eta)$ and so

$$\lim_{N\to\infty} |S(N^{1/(1-\eta)})| = \lim_{N\to\infty} |S(N)| = 0.$$

□

Combining the corollary with the estimate for $|E(N)|$ above and using (5.29), we get

$$|U \setminus A(N)| \leqslant |E(N)| + |S(N)| \to 0 \tag{5.30}$$

as $N \to \infty$, *i.e.*, $A(N)$ satisfies (5.9).

Next let H be a hypercube with $\ell(H) = \rho(N)$. Suppose that $u \in (H/2) \cap A(N)$. Then there exists a pair $(p, \mathbf{q})$ satisfying (5.28) and $N^{1-\eta} \leqslant |\mathbf{q}| \leqslant N$. For N sufficiently large, the hyperplane $\Pi_{p,\mathbf{q}}$ will be far from tangential to M_U. Hence

$$\mathrm{dist}_\infty(u, R_{p,\mathbf{q}}) \asymp \mathrm{dist}_\infty(\theta(u), \theta(R_{p,\mathbf{q}})) \asymp \mathrm{dist}_2(\theta(u), \Pi_{p,\mathbf{q}}) \asymp \frac{|\mathbf{q}\cdot\theta(u) - p|}{|\mathbf{q}|_2}$$

and there exist a $c^* > 0$ such that

$$\mathrm{dist}_\infty(u, R_{p,\mathbf{q}}) \leqslant c^*\, N^{-n+k\eta}(\log N)^k |\mathbf{q}|^{-1} \leqslant c^*\, N^{-n-1+\eta(k+1)}(\log N)^k$$

and a $c_* > 0$ such that

$$\mathrm{dist}_\infty(u, R_{p,\mathbf{q}}) \geqslant c_* |\mathbf{q}\cdot\theta(u) - p|\, |\mathbf{q}|^{-1}. \tag{5.31}$$

We show that the other ubiquity properties (5.10) and (5.11) hold for the family of resonant sets $\{R_{p,\mathbf{q}}\}$ where $\lfloor(p, \mathbf{q})\rfloor = |\mathbf{q}|$ when we choose

$$\rho(N) = 4c^* N^{-n-1+(k+1)\eta}(\log N)^k. \tag{5.32}$$

First, since $\mathrm{dist}_\infty(u, R_{p,\mathbf{q}}) \leqslant \ell(H)/4$, the set $R_{p,\mathbf{q}}$ meets the hypercube H substantially and

$$|H \cap B(R_{p,\mathbf{q}}\,; \delta)| \gg \ell(H)^{m-1}\delta.$$

Secondly since the hyperplane $\Pi_{p,\mathbf{q}}$ meets M_U far from tangentially, it meets M_U just once and so by the geometry, any hypercube H' with $\ell(H') \leqslant \rho(N)$ satisfies

$$|H' \cap H \cap B(R_{p,\mathbf{q}}\,; \delta)| \ll \ell(H')^{m-1} \min\{\delta, \ell(H')\} \ll \ell(H')^{m-1}\delta.$$

Thus the family $\{R_{p,\mathbf{q}} \colon \mathbf{q} \in \mathbb{Z}^n \setminus \{0\}, p \in \mathbb{Z}\}$ is ubiquitous with respect to ρ. Hence by Theorem 5.6, for any $\widetilde{\psi} \colon \mathbb{N} \to \mathbb{R}^+$,

$$\dim \Lambda(\{R_{p,\mathbf{q}}\}, \widetilde{\psi}) \geqslant m - 1 + \gamma,$$

where $\Lambda(\{R_{p,\mathbf{q}}\}, \widetilde{\psi})$ is the set of points u in U satisfying

$$\operatorname{dist}(u, R_{p,\mathbf{q}}) \leqslant \widetilde{\psi}(\lfloor(p, \mathbf{q})\rfloor) = \widetilde{\psi}(|\mathbf{q}|)$$

for infinitely many $p, \mathbf{q}$ and where $\gamma = \limsup_{N\to\infty} \log \rho(N)/\log \widetilde{\psi}(N)$.

Choose $\widetilde{\psi}(r) = \psi(r)/(c_* r)$. If $\operatorname{dist}(u, R_{p,\mathbf{q}}) < \widetilde{\psi}(|\mathbf{q}|)$, then $|\mathbf{q}\cdot\theta(u) - p| < \psi(|\mathbf{q}|)$ by (5.31). Therefore $\Lambda(\{R_{p,\mathbf{q}}\}, \widetilde{\psi}) \subseteq L(\psi)$. Hence by Theorem 3.5,

$$\dim L(\psi) \geqslant \dim \Lambda(\{R_{p,\mathbf{q}}\}, \widetilde{\psi}) \geqslant m - 1 + \gamma,$$

where since $\rho(N) \asymp N^{-n-1+\eta(k+1)}(\log N)^k$ and λ is the lower order of $1/\psi$,

$$\gamma = \limsup_{N\to\infty} \frac{\log \rho(N)}{\log(\psi(N)/N)} = \frac{n+1-\eta(k+1)}{\lambda+1}.$$

But η is an arbitrary positive number and U is a Monge domain and so

$$\dim \mathscr{L}(M\,;\psi) \geqslant \dim L(\psi) \geqslant m - 1 + \frac{n+1}{\lambda+1} \tag{5.33}$$

for $\lambda \geqslant n$.

By [139], a smooth m-dimensional manifold embedded in $\mathbb{R}^n$ and non-degenerate almost everywhere is extremal. Hence (5.33) holds for such manifolds and so in particular for manifolds which are 2-curved almost everywhere. Combining this with Lemma 4.7 gives the following result.

THEOREM 5.14. *Let M be a C^3 m-dimensional manifold embedded in $\mathbb{R}^n$, where $m \geqslant 2$. Let M_0, the set of points $x \in M$ at which M is not 2-curved, have Hausdorff dimension at most $m-1$. Then for any decreasing function $\psi\colon \mathbb{N} \to \mathbb{R}^+$,*

$$\dim \mathscr{L}(M\,;\psi) = m - 1 + \frac{n+1}{\lambda+1},$$

where λ is the lower order of $1/\psi$.

5.4. Simultaneous Diophantine approximation on manifolds

As mentioned in §4.4.1, determining the Hausdorff dimension of the set $\mathscr{S}_v(M)$ of simultaneously v-approximable points on manifolds can be more difficult than the dual case. In the case when M is the circle $\mathbb{S}^1$ and $v > 1$, the denominator q in (4.12) is part of the Pythagorean triple (p, r, q) (see §4.4.1). Melnichuk [161] used regular systems to obtain the lower bound

$$\dim \mathscr{S}_v(\mathbb{S}^1) \geqslant \frac{1}{2(v+1)}$$

for $v > 1$. Ubiquity can also be used and there is reason to suppose that in fact the lower bound coincides with the upper bound obtained in (4.14).

Exponential sums can also give estimates from below for the Hausdorff dimension of the set $\mathscr{S}_v(M)$ of simultaneously v-approximable points. We begin with some general considerations and then specialise to obtain a lower bound for the

Hausdorff dimension of $\mathscr{S}_v(M)$ where $v < 1/3$ and M is a product of planar curves with non-zero curvature except on a set of Hausdorff dimension 0. As in the case of the upper estimate, the argument involves the distribution of rational points $(\mathbf{p}/q, \mathbf{r}/q) \in U \times \mathbb{R}^k$ near the manifold M.

Recall that $\mathscr{S}(M;\psi)$ is the set of ξ in a smooth manifold M for which

$$|\langle q\,\xi\rangle| = |\langle(q\xi_1, \dots, q\xi_n)\rangle| < \psi(q)$$

holds for infinitely many positive integers q. To determine the Hausdorff dimension of $\mathscr{S}(M;\psi)$, it suffices to consider the set

$$S(\psi) = \{u \in U\colon |\langle q(u, \varphi(u))\rangle| < \psi(q) \text{ for infinitely many } q \in \mathbb{N}\},$$

where U is a suitably chosen hypercube in $\mathbb{R}^m$ and $\varphi\colon U \to \mathbb{R}^k$ is a Monge ordinate function satisfying (see §1.4.2)

$$\max\{\sup\{|\partial\varphi(u)/\partial u_i|\colon u \in U\}\colon i = 1, \dots, m\} = K < \infty \tag{5.34}$$

for a constant $K \geqslant 0$. For the moment we need only restrict M to being C^1.

As in §4.4 the number of integer vectors $\mathbf{p} \in qU = \{q\,u\colon u \in U\}$ satisfying $|\langle q\,\varphi(\mathbf{p}/q)\rangle| < \delta$ can be used to give an estimate for the Hausdorff dimension of $\mathscr{S}_v(M)$. The natural approach of expressing the number as a Fourier series and trying to show that the first term in the expansion dominates the remainder cannot work without modification since the Fourier series is not absolutely convergent. Nevertheless there are approximations in which the coefficients decay sufficiently rapidly to give absolute convergence. For simplicity in this section we assume (as in §4.4), that $U = [0,\, 1]^m$. We begin with the case $m = 1$.

5.4.1. Rational points near a curve. Given $\delta > 0$, the characteristic function of the set $\bigcup_{p\in\mathbb{Z}}\{\xi \in \mathbb{R}\colon |\xi - p| < \delta\}$ will be written $\chi_\delta\colon \mathbb{R} \to \mathbb{R}$, so that

$$\chi_\delta(t) = \begin{cases} 1, & \text{if } \|t\| < \delta, \\ 0, & \text{if } \|t\| \geqslant \delta. \end{cases}$$

The function χ_δ is 1-periodic, with Fourier series representation

$$\chi_\delta(t) \sim \sum_{j\in\mathbb{Z}} a_j e^{2\pi i t}. \tag{5.35}$$

For each $q \in \mathbb{N}$, let $\mathcal{N}(\varphi, q; \delta)$ be the number of points p/q in a given interval $(\alpha, \beta) \subset [0, 1]$ for which $\varphi(p/q)$ is within δ/q of a point r/q. Then

$$\mathcal{N}(\varphi, q; \delta) = \sum_{p\in(q\alpha, q\beta)} \chi_\delta(q\varphi(p/q)) \tag{5.36}$$

(the dependence on (α, β) is suppressed).

5.4.2. The Vinogradov tumbler functions. In order to overcome the conditional convergence of the Fourier series for χ_δ, Vinogradov's tumbler functions χ_δ^+, χ_δ^-, which are continuous piece-wise linear approximations to χ_δ, are introduced [227]. Let $\delta < 1/4$ and $0 < \eta < \delta$. Then χ_δ has a continuous approximation $\chi_\delta^+ : \mathbb{R} \to \mathbb{R}$ defined by

$$\chi_\delta^+(t) = \begin{cases} 1, & \text{when } \|t\| \leqslant \delta, \\ 1 - (\|t\| - \delta)/\eta, & \text{when } \delta < \langle t \rangle < \delta + \eta, \\ 1 + (\|t\| - \delta)/\eta, & \text{when } \delta - \eta < \langle t \rangle < \delta, \\ 0, & \text{if } \|t\| \geqslant \delta + \eta \end{cases}$$

(see [210, p. 53], where χ_δ^+ is termed an (upper) 'smoothed characteristic function', and Chapter 4, Lemma 2 of [59]).
Clearly $\chi_\delta^+(t) \geqslant \chi(t)$ for all $t \in \mathbb{R}$. The function χ_δ^+ has period 1 and absolutely convergent Fourier series

$$\chi_\delta^+(t) = \sum_{j \in \mathbb{Z}} a_j^+ \, e^{2\pi i j t},$$

where $a_j^+ = \int_0^1 \chi_\delta^+(t) e^{-2\pi i j t}\, dt$, $j \in \mathbb{Z}$, so that

$$a_0^+ = 2\delta + \eta, \; |a_j^+| \ll \min\{\delta + \eta, |j|^{-1}, \eta^{-1}|j|^{-2}\}, \; j \neq 0.$$

The lower approximation χ_δ^- for χ_δ is constructed analogously, giving

$$\chi_\delta^-(t) = \sum_{j \in \mathbb{Z}} a_j^- \, e^{2\pi i j t},$$

where

$$a_0^- = 2\delta - \eta, \quad |a_j^-| \ll \min\{\delta - \eta, |j|^{-1}, \eta^{-1}|j|^{-2}\}, \; j \neq 0. \tag{5.37}$$

5.4.3. Exponential sum estimates. It follows from the definition (5.36) of $\mathcal{N}(\varphi, q; \delta)$ that

$$\sum_{p \in q(\alpha,\beta)} \chi_\delta^-(q\varphi(p/q)) \leqslant \mathcal{N}(\varphi, q; \delta) \leqslant \sum_{p \in q(\alpha,\beta)} \chi_\delta^+(q\varphi(p/q)) \tag{5.38}$$

and, on substituting the Fourier series and interchanging the order of summation, that

$$(2\delta - \eta) q(\beta - \alpha) + R^-(q, \delta) \leqslant \mathcal{N}(\varphi, q; \delta) \leqslant (2\delta + \eta) q(\beta - \alpha) + R^+(q, \delta),$$

where

$$R^\pm(q, \delta) = \sum_{p \in q(\alpha,\beta)} \sum_{j \neq 0} a_j^\pm \, e^{2\pi i j q \varphi(p/q)} + O(1) = \sum_{j \neq 0} a_j^\pm \, E(q, j) + O(1)$$

and $E(q, j) = \sum_{p \in q(\alpha,\beta)} e^{2\pi i j q \varphi(p/q)}$. But by Lemma 4.10 and since $\varphi''(p/q) \asymp 1$,

$$\left| \sum_{p=0}^{q} e^{2\pi i j q \varphi(p/q)} \right| \ll \left(q \left(\frac{|j|}{q} \right) + 1 \right) \left(\frac{q}{|j|} \right)^{1/2} \ll (q|j|)^{1/2} + (q/|j|)^{1/2}.$$

Then

$$|E(q,j)| \ll \sum_{p\in(q\alpha,q\beta)} e^{2\pi i j q\varphi(p/q)} \ll \left(\frac{|j|}{q}q(\beta-\alpha)+1\right)\left(\frac{q}{|j|}\right)^{1/2}$$
$$\ll (|j|\,q)^{1/2}(\beta-\alpha)+(q/|j|)^{1/2} \ll q^{1/2}\left(|j|^{1/2}(\beta-\alpha)+|j|^{-1/2}\right).$$

Hence by (5.37),

$$\begin{aligned} R^-(q,\delta) &= \sum_{j\neq 0} a_j^- E(q,j) \\ &\ll q^{1/2} \sum_{1\leqslant|j|\leqslant(\delta-\eta)^{-1}} (\delta-\eta)\left(|j|^{1/2}(\beta-\alpha)+|j|^{-1/2}\right) \\ &\quad + q^{1/2} \sum_{(\delta-\eta)^{-1}\leqslant|j|\leqslant\eta^{-1}} |j|^{-1}\left(|j|^{1/2}(\beta-\alpha)+|j|^{-1/2}\right) \\ &\quad + q^{1/2} \sum_{|j|\geqslant\eta^{-1}} |j|^{-2}\eta^{-1}\left(|j|^{1/2}(\beta-\alpha)+|j|^{-1/2}\right). \end{aligned}$$

Let $\varepsilon>0$, $\delta=(2q^v)^{-1}$ and take $\eta=q^{-v-\varepsilon}/2$. Then the first sum is

$$\begin{aligned} &\ll q^{1/2} \sum_{1\leqslant|j|\ll q^v} \left(q^{-v}|j|^{1/2}(\beta-\alpha)+q^{-v}|j|^{-1/2}\right) \\ &\ll q^{1/2-v}(\beta-\alpha)\sum_{1\leqslant j\ll q^v} j^{1/2} + q^{1/2-v}\sum_{1\leqslant j\ll q^v} j^{-1/2} \\ &\ll q^{(1+v)/2}(\beta-\alpha)+q^{(1-v)/2}. \end{aligned}$$

The other sums can be shown to be $\ll q^{(1+v+\varepsilon)/2}(\beta-\alpha)+q^{(1-v)/2}$. Combining these estimates we get that

$$|R^-(q,q^{-v}/2)| \ll q^{(1+v+\varepsilon)/2}(\beta-\alpha)+q^{(1-v)/2}.$$

For convenience let $\mathcal{N}_v(\varphi)=\mathcal{N}(\varphi,q;q^{-v}/2)$. By (5.38),

$$\mathcal{N}_v(\varphi) \gg q^{1-v}(\beta-\alpha)+R^-(q;q^{-v}/2), \tag{5.39}$$

providing $1-v>(1+v+\varepsilon)/2$, and $1-v-t>(1-v)/2$, where $q^{-t}=\beta-\alpha$. If $v<1/3$ and $t<(1-v)/2$, then $\mathcal{N}_v(\varphi)>0$ and so *any interval* (α,β) *of length* $q^{-(1-v)/2+\varepsilon_0}$ *for* $\varepsilon_0>0$, *contains a rational* p/q *with* $|\langle q\varphi(p/q)\rangle|<(2q^v)^{-1}$. In fact $\mathcal{N}_v(\varphi)\asymp q^{1-v}(\beta-\alpha)$.

5.4.4. Rational vectors in $[0,1]^m$**.** We now consider higher dimensions and take $M=\prod_{i=1}^m \Gamma_i$, the Cartesian product of m planar curves

$$\{(u_i,\varphi_i(u_i))\colon u_i\in[0,1]\},$$

so that $k=m$ and the set S_v consists of points $u=(u_1,\dots,u_m)\in[0,1]^m$ such that the system of inequalities

$$\max\{\|qu_i\|,\|q\varphi_i(u_i)\|\colon i=1,\dots,m\}<q^{-v} \tag{5.40}$$

holds for infinitely many positive integers q. The number $\mathcal{N}_v(q)$ of solutions of

$$|\langle q(\varphi_1(p_1/q), \dots, \varphi_m(p_m/q))\rangle| < q^{-v}/2$$

is given by

$$\mathcal{N}_v(q) = \prod_{i=1}^{m} \mathcal{N}_v(\varphi_i),$$

where $\mathcal{N}_v(\varphi)$ is the number of solutions of (5.40) when $m = 1$. Hence if $v < 1/3$ and $t = (1-v)/2 - \varepsilon_0 > 0$, then $\mathcal{N}_v(q) > 0$ from above.

Ubiquity is used to obtain a lower bound for the Hausdorff dimension of S_v. Cover $[0,1]^m$ by hypercubes C of sidelength $\ell(C) = q^{-(1-v)/2+\varepsilon_0}$ and choose, as we may by (5.39), a vector $\mathbf{p}/q$ in each C and $\mathbf{r} \in \mathbb{Z}^m$ such that

$$\max\{|\varphi_i(p_i/q) - r_i/q| : i = 1, \dots, m\} < (2q^{v+1})^{-1}. \tag{5.41}$$

The union of such points $\mathbf{p}/q$ can be regarded as a resonant set R_q. The family $\mathscr{R} = \{R_q : q \in \mathbb{N}\}$ is trivially ubiquitous with respect to the function $\rho(N) = N^{-(1-v)/2+\varepsilon_0}$ by taking $A(N) = [0,1]^m$.

The Taylor series for φ about $\mathbf{p}/q$ gives

$$\varphi(u) = \varphi(\mathbf{p}/q) + \mathcal{R},$$

where $|\mathcal{R}| \leqslant c|u - \mathbf{p}/q|$ for some $c > 0$ (as in §4.4.2). The set of points $u \in [0,1]^m$ such that

$$\left|u - \frac{\mathbf{p}}{q}\right| = \max\left\{\left|u_i - \frac{p_i}{q}\right| : i = 1, \dots, m\right\} < \frac{1}{2cq^{v+1}}$$

for infinitely many $\mathbf{p}/q \in \mathscr{R}$ will be denoted by $\Lambda_v(\mathscr{R})$. Then $\Lambda_v(\mathscr{R}) \subseteq S_v$ since if $u \in \Lambda_v(\mathscr{R})$, then by (5.41) there exists a $q \in \mathbb{N}$ such that $\mathbf{p}/q \in \mathscr{R}$ for which

$$\max\left\{\left|\varphi_i\left(\frac{p_i}{q}\right) - \frac{r_i}{q}\right| : i = 1, \dots, m\right\} < \frac{1}{2q^{v+1}}.$$

Hence for each $i = 1, \dots, m$,

$$\begin{aligned}\left|\varphi_i(u_i) - \frac{r_i}{q}\right| &\leqslant \left|\varphi(u_i) - \varphi(\frac{p_i}{q})\right| + \left|\varphi_i(\frac{p_i}{q}) - \frac{r_i}{q}\right| \\ &\leqslant c\left|u_i - \frac{p_i}{q}\right| + \frac{1}{2q^{v+1}} \leqslant \frac{1}{2q^{v+1}} + \frac{1}{2q^{v+1}} = \frac{1}{q^{v+1}}\end{aligned}$$

by Taylor's theorem. Since points in $[0,1]^m$ have codimension m, Theorem 5.6 above implies that

$$\dim S_v \geqslant \dim \Lambda_v(\mathscr{R}) \geqslant 0 + m \limsup_{N\to\infty} \frac{\log \rho(N)}{\log N^{-(v+1)}} = m\frac{1 - v - 2\varepsilon_0}{2(1+v)}.$$

But $\varepsilon_0 > 0$ is arbitrary and the bound is independent of the local coordinate system, so that

$$\dim \mathscr{S}_v\left(\prod_{i=1}^{m}\Gamma_i\right) = \dim S_v \geqslant \frac{m(1-v)}{2(1+v)}.$$

Combining this with the upper estimates for $\dim S_v$ obtained in §4.4.2, we get

$$\frac{m(1-v)}{2(1+v)} \leqslant \dim \mathscr{S}_v\left(\prod_{i=1}^{m}\Gamma_i\right) \leqslant \frac{1+(1-v)m}{1+v}$$

when $v < 1/3$ (see (4.22) for an upper estimate when $v \geqslant 1/3$).

5.5. Notes

§5.1 A lower bound for the Hausdorff dimension of systems of linear forms in terms of simultaneous Diophantine approximation with a general type of approximation function is given in [80].
§5.2 B. P. Rynne [193] has introduced the notion of 'weak regularity' which is a little more general than regular systems.

Theorem 5.1 can be extended to countable families of regular systems [13].

In order to study Diophantine approximation in hyperbolic space, Melián and Pestana [158] extended regular systems to systems of balls in $\mathbb{R}^n$. A countable collection $\mathcal{B}$ of Euclidean balls $B_i = B(a_{(i)}, r_i) = \{x \in \mathbb{R}^n : |x - a_{(i)}|_2 < r_i\}$ is said to be a *well distributed system of balls with constant* c if for every ball B in $\mathbb{R}^n$, there exists a positive number $K(B)$ for which given any K with $K \geqslant K(B)$, there is a subcollection $\mathcal{B}(K, B) \subset \mathcal{B}$ satisfying

(1) $a_{(i)} \in B$ and $r_i \geqslant K^{-1}$ for all $B_i \in \mathcal{B}(K, B)$,
(2) for all distinct B_i, $B_j \in \mathcal{B}(K, B)$, $|a_{(i)} - a_{(j)}|_2 > \min\{r_i, r_j\}$,
(3) the number of $\mathcal{B}(K, B)$ is at least $cK^n|B|$.

Balls in $\mathbb{C}$ of radius $1/|w|^2$ centred at z/w where z, w are Gaussian integers with $w \neq 0$ form a well distributed system. The Hausdorff dimension of directions of geodesics which emanate from a point and which approach cusps at certain rates (see Chapter 7 for more details) is obtained and the Jarník-Besicovitch theorem and its complex version deduced.
§5.2.2 J. Levesley [149] used uniform distribution and ubiquity to obtain the inhomogeneous form of the general Jarník-Besicovitch theorem.
§5.3 Any norm equivalent to the supremum norm could be used in the definition.

The requirement that the sequence $A(N)$ in (5.9) approximates U in measure can be relaxed to the condition that it approximates a constant proportion of each subset of U; this is the condition for regular systems [193].

In [54], [83], [96], mean and variance ideas related to second moment arguments in [59, Chapter VII] and a repeated subdivision technique from [48] are used. The former relies on a variance associated with the distribution of resonant planes not being too large and the second has similarities with fractals which reproduce themselves in some sense at certain scales (see for example [100], [101]).

In [193], Rynne has shown that classes of sets with 'large intersection' constructed from the $\mathcal{M}_\infty^s$-dense sequences introduced by Falconer [99] can also be constructed using ubiquity.
§5.3.1 If when $\gamma > 1$, $\lim_{r\to\infty}(\psi(N_r)^\gamma/\rho(N_r)) = 0$, then the Hausdorff measure is infinite at the critical exponent [90]. This latitude in the choice of ψ (which also exists in regular systems) allows the inference that the Hausdorff measure is positive at the corresponding critical exponent and implies that the Hausdorff dimensions of the sets $\mathcal{K}'_v(\mathbb{R})$, $\mathcal{S}'_v(\mathbb{R}^n)$ and $\mathcal{L}'_v(\mathbb{R}^n)$ of exact exponent of approximation v are the same as those of $\mathscr{K}_v(\mathbb{R})$, $\mathscr{S}_v(\mathbb{R}^n)$ and $\mathscr{L}_v(\mathbb{R}^n)$ (see §3.5.6). The determination of Hausdorff measure of $\mathscr{S}_v(\mathbb{R}^n)$ and $\mathscr{L}_v(\mathbb{R}^n)$ at the critical exponent needs a deeper approach [79, 127]. In their generalisation [79] of [127] to systems of

linear forms, Dickinson and Velani use a Cantor construction to obtain the Hausdorff measure and deduce that it is infinite at the critical exponent for $\mathscr{S}_v(\mathbb{R}^n)$ and $\mathscr{L}_v(\mathbb{R}^n)$.

When $\gamma = 1$, the Hausdorff measure is comparable to Lebesgue measure and can be finite.

Theorem 5.10 implies that the Hausdorff dimension of the set W_v of matrices $A \in \mathbb{R}^{mn}$ for which $|\langle \mathbf{q}A \rangle| < |\mathbf{q}|^{-v}$ for infinitely many $\mathbf{q} \in \mathbb{Z}^m$ is $m - 1 + (m+n)/(v+1)$ for $v \geqslant m/n$ and mn otherwise. Thus $\dim W_v$ is a continuous function of v, whence

$$\dim \mathscr{L}_v(\mathbb{R}^n) = \dim \mathcal{L}_v(\mathbb{R}^n) \ (= \dim \mathcal{L}'_v(\mathbb{R}^n)).$$

CHAPTER 6

Diophantine approximation over the p-adic field

6.1. Introduction to p-adic numbers

A brief account of p-adic numbers is given for completeness and to set notation (more detailed accounts can be found in the books by Cassels [61] and W. H. Schikof [195]). Throughout this chapter, p will be a prime number. Every non-zero rational number a can be expressed uniquely in the form $a = p^m a'$ where the numerator and denominator of a' are prime to p and $m \in \mathbb{Z}$ (m is said to be a rational integer). The *p-adic valuation* $|a|_p$ of a is defined to be

$$|a|_p = \begin{cases} 0 & \text{if} \quad a = 0, \\ p^{-m} & \text{if} \quad a \neq 0. \end{cases}$$

For any a and b in $\mathbb{Q}$, $|ab|_p = |a|_p |b|_p$,

$$|a + b|_p \leqslant \max\{|a|_p, |b|_p\}$$

and if $|a|_p \neq |b|_p$, then

$$|a + b|_p = \max\{|a|_p, |b|_p\}, \tag{6.1}$$

making this valuation an ultrametric. It is useful to note that for $a \neq 0$,

$$|a|_p \geqslant 1/|a|,$$

where $|a|$ denotes the usual absolute value of a. Just as the completion of the field $\mathbb{Q}$ with respect to the usual metric is the field $\mathbb{R}$, the completions of $\mathbb{Q}$ with respect to the p-adic metrics $|\cdot|_p$ give rise to the fields $\mathbb{Q}_p$; these are not isomorphic for different p. The elements are p-adic numbers. Each element ξ of $\mathbb{Q}_p$ has a unique representation

$$\xi = \sum_{r=m}^{\infty} c_r p^r,$$

where $m \in \mathbb{Z}$ and the coefficients c_r are rational integers $0 \leqslant c_r \leqslant p - 1$. If $c_m \neq 0$ then $|\xi|_p = p^{-m}$; if m is non-negative, $|\xi|_p \leqslant 1$ and ξ is called a *p-adic integer*; rational integers are p-adic integers. The subset of p-adic integers forms a ring denoted $\mathbb{Z}_p$.

Being a locally compact abelian group under addition, $\mathbb{Q}_p$ has a translationally invariant Haar measure which we denote by μ and which is unique up to a multiplicative constant. The ring $\mathbb{Z}_p$ is compact and the constant is usually fixed so that $\mu(\mathbb{Z}_p) = 1$; $\mathbb{Z}_p$ can be regarded as an analogue of the closed interval $[-1, 1]$ in

$\mathbb{R}$. A great advantage is that it is possible to construct such a measure directly. First, the Haar measure $\mu(D(\xi,p^{-k}))$ of the disc

$$D(\xi,p^{-k})=\{\alpha\in\mathbb{Q}_p\colon |\alpha-\xi|_p\leqslant p^{-k}\}$$

is given by $\mu(D(\xi,p^{-k}))=p^{-k}$. Now since the p-adic metric is an ultrametric, each point of a disc may be considered as its centre. It follows that if two discs intersect then one contains the other, whence closed sets are also open and vice versa.

For any finite union of discs $\bigcup_{i\in J}D_i$ we can choose a subset $\{D_i\colon i\in J'\}$, where $J'\subset J$, of discs which are disjoint. Thus by disjointness $\bigcup_{i\in J}D_i=\bigcup_{i\in J'}D_i$, and the Haar measure of $\bigcup_{i\in J}D_i$ is $\sum_{i\in J'}\mu(D_i)$. Since the set $\mathcal{D}$ of all discs is closed with respect to finite unions and intersections, the measure μ can be extended to the smallest σ-field $\mathcal{F}$ over $\mathcal{D}$ by defining for each $T\in\mathcal{F}$,

$$\mu(T)=\inf\sum_{i\in J'}\mu(D_i),$$

where D_i, $i\in J'$ are disjoint, J' is a countable set and where the infimum is taken over all covers of T by unions of discs $\bigcup_{i\in J'}D_i$.

The p-adic norm on the vector space $\mathbb{Q}_p^n$ is defined by

$$|\xi|_p=\max\{|\xi_1|_p,\dots,|\xi_n|_p\}$$

for each $\xi=(\xi_1,\dots,\xi_n)\in\mathbb{Q}_p^n$; for each $\xi'=(\xi_1',\dots,\xi_n')\in\mathbb{Q}_p^n$

$$|\xi+\xi'|_p\leqslant\max\{|\xi|_p,|\xi'|_p\},\ |\xi\cdot\xi'|_p=|\sum_{k=1}^n\xi_k\xi_k'|_p\leqslant\max_k\{|\xi_k\xi_k'|_p\}\leqslant|\xi|_p|\xi'|_p$$

and for any $\alpha\in\mathbb{Q}_p$, $|\alpha\,\xi|_p=|\alpha|_p|\xi|_p$. Rational integer vectors such as $\mathbf{q}$, $\mathbf{r}$ will continue to be written in bold; general points in $\mathbb{Z}_p^n$ will not. Recall that for $\mathbf{q}=(q_0,q_1,\dots,q_n)\in\mathbb{Z}^{n+1}$, we write $|\mathbf{q}|=\max\{|q_i|\colon 0\leqslant i\leqslant n\}$. Haar measure can be extended to $\mathbb{Q}_p^n$ in a natural way by defining $\mu(D(\xi,p^{-t}))=p^{-t}$ for the disc $D(\xi,p^{-t})=\{\alpha\in\mathbb{Q}_p^n\colon|\alpha-\xi|_p\leqslant p^{-k}\}$ [151].

The definition of Hausdorff measure on $\mathbb{Q}_p^n$ in the same as on $\mathbb{R}^n$. Given $s>0$ and a subset E of $\mathbb{Q}_p^n$, the Hausdorff s-measure is given by

$$\mathcal{H}^s(E)=\sup_{\delta>0}\{\inf\sum_j\operatorname{diam}D_j\}$$

where the infimum is over all covers of E by discs D_j of diameter (in the p-adic metric) at most δ. The Hausdorff dimension of E is defined by

$$\dim E=\inf\{s\colon\mathcal{H}^s(E)=0\}.$$

It is a well known fact that a valuation defined on a field F can be extended to a valuation on a given extension K of degree k over F [61]. If F is complete with respect to the given valuation, then the finite extension K is also complete with respect to the extension of the valuation on F defined by the equation

$$|\xi|_p = |\mathrm{Nm}(\xi)|_p^{1/k}$$

for each $\xi \in K$, where $\mathrm{Nm}(\xi) = \xi\,\xi^{(2)} \dots \xi^{(k)}$, the product of the algebraic conjugates of ξ, is the algebraic norm of the element ξ in F.

6.2. Diophantine approximation in $\mathbb{Q}_p$

The study of p-adic Diophantine approximation was begun by Mahler [153] and Jarník [128, 129]; more general results for homogeneous and inhomogeneous approximation may be found in the monograph of E. Lutz [151]. In order to conform with the earlier part of this book, the notation adopted here differs from that theirs. The first result is from [153].

LEMMA 6.1. *Let* $a_{ij} \in \mathbb{R}$, $i, j = 0, \dots, n$, *and suppose that* $\det(a_{ij}) \neq 0$. *Let* $\alpha_i \in \mathbb{Z}_p$ *for* $i = 0, 1, \dots, n$. *Then for any* $\beta, \beta_0, \dots, \beta_n \in (0, 1)$ *which satisfy* $\beta\beta_0\beta_1 \dots \beta_n \geqslant p|\det(a_{ij})|$, *there exists a non-zero vector* $\mathbf{q} = (q_0, q_1, \dots, q_n)$ *in* $\mathbb{Z}^{n+1}$ *such that the system of* $n+1$ *inequalities*

$$\left|\sum_{i=0}^{n} q_i \alpha_i\right|_p < \beta, \quad \left|\sum_{i=0}^{n} q_i a_{ij}\right| \leqslant \beta_j, \quad j = 0, 1, \dots, n,$$

holds.

The linear forms in the statement of the lemma can be written more concisely in matrix form as ξA where $\xi = (\xi_0, \xi_1, \dots, \xi_n)$ and $A = (a_{ij})$. The matrix $A = (a_{ij})$ can be treated as a point in $\mathbb{Z}_p^{mn}$. The next results follow from Lemma 6.1.

LEMMA 6.2. *Let* $\xi = (\xi_0, \dots, \xi_n) \in \mathbb{Z}_p^{n+1}$. *Then for each rational integer* $N \geqslant 2$, *there exists a non-zero vector* $\mathbf{q}$ *in* $\mathbb{Z}^{n+1}$ *with* $|\mathbf{q}| \leqslant N$ *such that*

$$\left|\sum_{i=0}^{n} q_i \xi_i\right|_p = |\mathbf{q} \cdot \xi|_p < pN^{-n-1}.$$

This lemma is the special case of Lemma 6.1 when the matrix $A = (a_{ij})$ is diagonal, $\beta_i = N$, $i = 0, 1, \dots, n$, and $\beta = pN^{-n-1}$. It is also easily proved using Dirichlet's pigeonhole principle and it implies that $\omega(\xi) = \omega_n(\xi)$, the supremum of the set of those $w > 0$ for which there exist infinitely many vectors $(q_0, q_1, \dots, q_n)$ in $\mathbb{Z}^{n+1}$ satisfying the inequality

$$|q_n \xi^n + \dots + q_1 \xi + q_0|_p < |\mathbf{q}|^{-w},$$

is at least $n+1$ for each $n = 1, 2, \dots$

The next theorem is a p-adic analogue of an inhomogeneous Khintchine-Groshev theorem for a set of m inhomogeneous linear forms over $\mathbb{Z}_p$ in $(m+n)$ variables (see Theorems 4.11 and 4.12 in [151]). We consider the canonical system given by

$$\sum_{j=1}^{n} q_j a_{ij} + r_i + b_i, \; i = 1, \ldots, m,$$

which will be written $\mathbf{q}A + \mathbf{r} + b$; the homogeneous case corresponds to $b = 0$.

THEOREM 6.3. *Let $\psi\colon \mathbb{N} \to \mathbb{R}^+$ be a 'sufficiently regular' function. Then the number of solutions $(\mathbf{q}, \mathbf{r}) \in \mathbb{Z}^m \times \mathbb{Z}^n$ of the inequality*

$$|\mathbf{q}A + \mathbf{r} + b|_p \leqslant \psi(|(\mathbf{q}, \mathbf{r})|)$$

is finite or infinite for almost all $(A, b) \in \mathbb{Z}_p^{mn+n}$ according as the series

$$\sum_{r=1}^{\infty} r^{m+n-1} \psi(r)^m$$

converges or diverges.

The expression 'sufficiently regular' means different things in different cases of the theorem. In the divergence case it means that the approximation function ψ must also satisfy $r\psi(r) \to 0$ as $r \to \infty$. In the case of convergence it means that the decreasing positive function ψ satisfies $r\psi(r) < 1/2$ for $r \geqslant 1$, and for any $c' > 0$ there exists a $c > 0$ such that for sufficiently large r we have $\psi(c'r) \leqslant c\psi(r)$. The function $\psi(r) = cr^{-v}$ obeys these conditions for positive constants c and v. The theorem implies that the number of solutions of the inequality

$$|\mathbf{q}A + \mathbf{r} + b|_p \leqslant c|(\mathbf{q}, \mathbf{r})|^{-v}$$

is finite or infinite for almost all $(A, b) \in \mathbb{Z}_p^{mn} \times \mathbb{Z}_p^n$ according as $v > (m+n)/m$ or $v \leqslant (m+n)/m$. There is a p-adic counterpart to the general form of the Jarník-Besicovitch theorem [1, 77].

6.3. Integral polynomials with small p-adic values

We now discuss the Hausdorff dimension of $\widetilde{\mathfrak{M}}_v$ $\widetilde{\mathscr{M}}_v$, the p-adic counterparts of $\mathfrak{M}_v$ and $\mathscr{M}_v$ introduced in Chapter 1. Thus $\widetilde{\mathfrak{M}}_v$ is the set of $\xi \in \mathbb{Q}_p$ such that

$$|P(\xi)|_p < h(P)^{-v}$$

for infinitely many integral polynomials of degree at most n and

$$\widetilde{\mathscr{M}}_v = \{\xi \in \mathbb{Q}_p \colon \omega(\xi) \leqslant v\} \quad \text{and} \quad \widetilde{\mathscr{M}}_v' = \{\xi \in \mathbb{Q}_p \colon \omega(\xi) = v\},$$

where $\omega(\xi) = \sup\{v \in \mathbb{R} \colon \xi \in \mathfrak{M}_v\}$.

Let $\widetilde{\mathcal{M}}_v$ consist of points ξ in the set $\widetilde{\mathscr{M}}_v$ for which the polynomials P are of degree n, irreducible and leading. Sprindžuk proved the p-adic version of Mahler's conjecture by showing that the sets $\widetilde{\mathfrak{M}}_v$ and $\widetilde{\mathscr{M}}_v$ are of Haar measure zero when $v > n+1$ [208, Chapter 2]; the Hausdorff dimension of these sets is now discussed.

We restrict ourselves without loss of generality to $\mathbb{Z}_p$, since if $|\xi|_p > 1$ we apply the transformation $\xi \mapsto \xi^{-1}$ which leaves the value of the Hausdorff dimension unchanged (see Corollary 3.8). I. L. Morotskaya [164] obtained the upper bound $\dim \widetilde{\mathscr{M}_v} \leqslant (n+1)/v$ for $v > n+1$; the complementary inequality is a consequence of [44]. This result is a p-adic analogue of [33] and has a similar proof on the same lines as those for integral polynomials discussed in Chapters 4 and 5, so complete details will be not given.

6.3.1. Upper bounds for Hausdorff dimension. The proof of the upper bound for arbitrary n is long and difficult but the basic idea is to construct economical covers. In this section we consider only the technically simple cases $n = 1$ and 2.

LEMMA 6.4. *When $n = 1$ and $v > 2$ we have* $\dim \widetilde{\mathscr{M}_v} < 2/v$.

PROOF. Let $\xi \in \widetilde{\mathscr{M}_v}$. Then for any $w \in (0, v)$, there exist $a_0, a_1 \in \mathbb{Z}$ such that

$$|a_0 + a_1\xi|_p < |(a_0, a_1)|^{-w}$$

and which can be taken to be coprime (recall that $|(a_0, a_1)| = \max\{|a_0|, |a_1|\}$). Moreover a_1 is coprime to p since otherwise by (6.1), $|a_0 + a_1\xi|_p = |a_0|_p = 1$. Thus $\xi \in \widetilde{\mathscr{M}_v} \cap \mathbb{Z}_p$ if and only if for any $w \in (0, v)$,

$$\left|\xi + \frac{a_0}{a_1}\right|_p < |(a_0, a_1)|^{-w} \tag{6.2}$$

for infinitely many rationals a_0/a_1. Hence $\widetilde{\mathscr{M}_v} \cap \mathbb{Z}_p$ has a natural cover

$$\mathscr{C} = \{D(a_0/a_1; |(a_0, a_1)|^{-w})\colon a_0 \in \mathbb{Z}, a_1 \in \mathbb{N}\}$$

consisting of discs. The s-length of $\mathscr{C}$ is at most

$$\sum_{a_0 \in \mathbb{Z}} \sum_{a_1=1}^{|a_0|} |a_0|^{-sw} + \sum_{a_1 \in \mathbb{N}} \sum_{|a_0| \leqslant a_1} |a_1|^{-sw} \ll \sum_{j \in \mathbb{N}} j^{-sw+1} < \infty$$

for $s > 2/w$ and so by the Haar measure version of Hausdorff-Cantelli lemma, $\dim \widetilde{\mathscr{M}_w} \leqslant 2/v$. $\square$

To construct the cover we have to anticipate a later result.

LEMMA 6.5. *When $n = 2$ and $v > 3$,* $\dim \widetilde{\mathscr{M}_v} \leqslant 3/v$.

PROOF. We consider the set $\mathfrak{M}_w$, where $3 < w < v$ of ξ, for which the inequality

$$|Q(\xi)|_p < h(Q)^{-w}$$

holds for infinitely many integer quadratics $Q(\xi) = a_2\xi^2 + a_1\xi + a_0$. We may suppose, as in §4.2, that these quadratics are irreducible and the height $h(Q) = N$;

in addition we take $|N|_p > p^{-2}$. Let α_1 be the nearest root of Q to ξ. Then Lemma 6.9 below implies that

$$|\xi - \alpha_1|_p \leqslant \frac{|Q(\xi)|_p}{|Q'(\alpha_1)|_p}.$$

Since $Q'(\alpha_1) = \Delta(Q)^{1/2}$ ($\Delta(Q)$ is the discriminant of Q), it follows that

$$|\xi - \alpha_1|_p \leqslant |Q(\xi)|_p |\Delta(Q)|_p^{-1/2} \leqslant N^{-w} |\Delta(Q)|_p^{-1/2}.$$

This gives a cover $\mathscr{C}$ for $\widetilde{\mathcal{M}}_w$ and we proceed to estimate its s-length $\ell^s(\mathscr{C})$ when $0 < \varepsilon < 1 - 3/w$ and $s = 3/w + \varepsilon$.

We begin by estimating the number $\mathcal{N}(j)$ of pairs of integers a_0, a_1 satisfying $\max\{|a_0|, |a_1|\} \leqslant a_2$ and the equation

$$\Delta(Q) = a_1^2 - 4a_0 N = j$$

for a fixed integer $j \neq 0$. It is shown in [208, p. 96, Lemma 12] that

$$\mathcal{N}(j) \ll (j, N)^{1/2} N^{\varepsilon},$$

where (j, N) is the greatest common divisor of j and N. Hence

$$\begin{aligned} \sum_{h(Q)=N} |\Delta(Q)|_p^{-s/2} &\leqslant \sum_{1 \leqslant j \ll N^2} \mathcal{N}(j) |j|_p^{-s/2} \ll \sum_{1 \leqslant j \ll N^2} N^{\varepsilon} (j, N)^{1/2} |j|_p^{-s/2} \\ &\leqslant \left(\sum_{1 \leqslant j \ll N^2} (j, N) \right)^{1/2} \left(\sum_{1 \leqslant j \ll N^2} |j|_p^{-s} \right)^{1/2} N^{\varepsilon} \end{aligned} \tag{6.3}$$

by the Cauchy-Schwarz-Bunyakovsky inequality and

$$\sum_{1 \leqslant j \ll N^2} (j, N) \ll \left(\sum_j \sum_{d|(j,N)} d \right) \ll \left(\sum_{d|N} d \frac{N^2}{d} \right) = N^2 \tau(N) \ll N^{2+\varepsilon},$$

where $\tau(N)$ is the number of divisors of N. In order to estimate the second factor let $\mathcal{N}_k$ be the number of j, $1 \leqslant j \ll N^2$, for which $|j|_p = p^{-k}$. Since $\mathcal{N}_k \ll N^2/p^k$ and since k satisfies $0 \leqslant k \ll \log N / \log p$,

$$\sum_{1 \leqslant j \ll N^2} |j|_p^{-s} \asymp \sum_{k \ll \log N} \mathcal{N}_k p^{ks} \ll \sum_{k \ll \log N} N^2 p^{k(3/w+\varepsilon-1)} \ll N^2,$$

since $w > 3$. Thus the s-length $\ell^s(\mathscr{C})$ for $\mathcal{M}_w$ satisfies

$$\ell^s(\mathscr{C}) \ll \sum_{N=1}^{\infty} \sum_{|a_0|, |a_1| \leqslant N} |\xi - \alpha_1|_p^s \ll \sum_{N=1}^{\infty} N^{-ws} \sum_{|a_0|, |a_1| \leqslant N} |\Delta(Q)|^{s/2}.$$

Hence using (6.3),

$$\ell^s(\mathscr{C}) \ll \sum_{N=1}^{\infty} N^{-w\left(\frac{3}{w}+\varepsilon\right)} N^{2+3\varepsilon/2} \ll \sum_{N=1}^{\infty} N^{-1-3\varepsilon/2} < \infty$$

and the result follows from the Hausdorff-Cantelli lemma. □

6.3.2. Lemmas on polynomials. Let $P(x) = a_n x^n + a_{n-1}x^{n-1} + \cdots + a_1 x + a_0$ be a polynomial over K, where K is a finite extension field of $\mathbb{Q}_p$. The set of irreducible primitive polynomials P, of degree $n \geqslant 2$ and with rational integral coefficients $a_0, a_1, \ldots, a_n$ satisfying

$$\max\{|a_0|, |a_1|, \ldots, |a_{n-1}|\} \leqslant a_n = N, \tag{6.4}$$

is denoted by $\mathfrak{I}_n(N)$ (see §2.4.1). Let N be fixed with $|N|_p > p^{-n}$. Let $\mathbb{Q}_p^*$ be the smallest field containing $\mathbb{Q}_p$ and all algebraic numbers. Since the set of all algebraic numbers is countable, it follows from the considerations in §6.1 that the p-adic valuation $\mathbb{Q}_p$ may be extended to a valuation on $\mathbb{Q}_p^*$. We shall use the notation $|\xi|_p$ for all $\xi \in \mathbb{Q}_p^*$.

LEMMA 6.6. *Each root α of a polynomial $P \in \mathfrak{I}_n(N)$ satisfies $|\alpha|_p < p$.*

PROOF. The algebraic p-adic number α_i is a root of the polynomial

$$P(x) = Nx^n + a_{n-1}x^{n-1} + \cdots + a_1 x + a_0$$

and since P is irreducible we have

$$|\alpha_i|_p = |\mathrm{Nm}\alpha_i|_p^{1/n} = |a_0/N|_p^{1/n} < (p^n)^{1/n} = p.$$

□

We order the roots $\alpha_1, \alpha_2, \ldots, \alpha_n$ of P (by changing indices if necessary) so that

$$|\alpha_1 - \alpha_2|_p \leqslant |\alpha_1 - \alpha_3|_p \leqslant \cdots \leqslant |\alpha_1 - \alpha_n|_p.$$

For each root α_i we define the set $\mathcal{T}(\alpha_i)$ to be those points ξ in $\mathbb{Q}_p$ which satisfy the condition $\min\{|\xi - \alpha_j|_p \colon j = 1, \ldots, n\} = |\xi - \alpha_i|_p$ for $j = 1, 2, \ldots, n$.

As in Chapter 2, let ε be a sufficiently small positive number and let $\varepsilon_1 = c(n)\,\varepsilon$, where the number $c(n)$ is sufficiently small. We introduce an integer T and real numbers μ_i by the equations

$$T = [\varepsilon_1^{-1}] + 1, \quad |\alpha_1 - \alpha_i|_p = N^{-\mu_i} \tag{6.5}$$

for $i = 2, 3, \ldots, n$. Let the integers $r_2, \ldots, r_n$ be defined by the inequalities

$$(r_i - 1)/T \leqslant \mu_i < r_i/T.$$

It follows from Lemma 6.6 and (6.5) that $r_i/T \geqslant 0$ for large N, whence $r_i \geqslant 0$. Let

$$\lambda_i = \frac{r_{i+1} + \cdots + r_n}{T} \tag{6.6}$$

for $i = 1, \ldots, n-1$. It is clear that $\mu_2 \geqslant \mu_3 \geqslant \cdots \geqslant \mu_n$ and therefore that $r_2 \geqslant r_3 \geqslant \cdots \geqslant r_n$ and $\lambda_2 \geqslant \lambda_3 \geqslant \cdots \geqslant \lambda_{n-1}$. For each polynomial $P \in \mathfrak{I}_n(N)$, the vector $\mathbf{r} = \mathbf{r}(P) = (r_2, \ldots, r_n)$ has non-negative integer components.

LEMMA 6.7. *The number of vectors* $\mathbf{r}$ *is finite and depends only on n and ε.*

This lemma is proved analogously to Lemma 2.5. The set of polynomials P in $\mathfrak{I}_n(N)$ which have the same vector $\mathbf{r}$ is denoted by $\mathfrak{I}_n(N, \mathbf{r})$.

LEMMA 6.8. *Let P be a reducible polynomial with integer coefficients and degree at most n and let δ be any positive number. Then for almost all $\xi \in \mathbb{Q}_p$ the inequality*

$$|P(\xi)|_p < h(P)^{-n-\delta}$$

has only a finite number of solutions.

The proof can be constructed by modifying Lemma 2.3 as Sprindžuk's p-adic theorem [208, p. 112] implies that only a finite number of integral polynomials with degree at most $n-1$ satisfy the inequality $|P(\xi)|_p < h(P)^{-n-\delta}$.

LEMMA 6.9. *Let $P \in \mathfrak{I}_n(N)$ and $\xi \in \mathcal{T}(\alpha_1)$. Then*

$$|\xi - \alpha_1|_p \leqslant \frac{|P(\xi)|_p}{|P'(\alpha_1)|_p}, \tag{6.7}$$

and for $2 \leqslant j \leqslant n$,

$$|\xi - \alpha_1|_p \leqslant \left(\frac{|P(\xi)|_p}{|P'(\alpha_1)|_p}|(\alpha_1 - \alpha_2)\cdots(\alpha_1 - \alpha_j)|_p\right)^{1/j}. \tag{6.8}$$

This lemma can be obtained in the same way as Lemma 2.7.

LEMMA 6.10. *Suppose $P \in \mathfrak{I}_n(N, \mathbf{r})$. Then for each $j = 1, \dots, n-1$*

$$|P^{(j)}(\alpha_1)|_p \leqslant N^{-\lambda_j + (n-j)\varepsilon_1}.$$

The proof of this is essentially the same as that of Lemma 2.8.

LEMMA 6.11. *Let $P \in \mathfrak{I}_n(N)$, $\xi \in \mathcal{T}(\alpha_1)$ and*

$$\left.\begin{array}{rcl} |P(\xi)|_p & < & N^{-n-1}, \\ |P'(\xi)|_p & < & (\log N)^{-\gamma}, \quad \gamma > 0. \end{array}\right\} \tag{6.9}$$

Then $|P'(\alpha_1)|_p \ll (\log N)^{-\gamma}$.

PROOF. From (6.8) with $j = n$ and the first inequality in (6.9), we have

$$|\xi - \alpha_1|_p \ll \left(\frac{|P(\xi)|_p}{|N|_p}\right)^{1/n} < pN^{-1-1/n}. \tag{6.10}$$

The Taylor series for the polynomial $P'(\xi)$ in the neighbourhood of the root α_1 is

$$P'(\xi) = P'(\alpha_1) + P''(\alpha_1)(\xi - \alpha_1) + \cdots + \frac{P^{(n)}(\alpha_1)}{(n-1)!}(\xi - \alpha_1)^{n-1}. \tag{6.11}$$

By Lemma 6.10, for $j \geqslant 2$

$$\left|\frac{P^{(j)}(\alpha_1)}{(j-1)!}(\xi - \alpha_1)^{j-1}\right|_p < \frac{N^{-\lambda_j + \varepsilon_1(n-1)}p^{j-1}N^{(-j-1)(1+1/n)}}{|(j-1)!|_p} < N^{-1}$$

for large N. Since $|P'(\xi)|_p < (\log N)^{-\gamma}$, the lemma must follow to avoid contradicting (6.11). $\square$

LEMMA 6.12. *Let $D(\xi, p^{-l})$ be some disc in $\mathbb{Q}_p$ and $B \subset D(\xi, p^{-l})$ be a measurable set of $\xi \in \mathbb{Q}_p$ with $\mu(B) \geqslant p^{-l}/k$ where $k \in \mathbb{N}$. Suppose that $|P(\xi)|_p < N^{-v}$ for each $x \in B$, where $v > 0$ and $\deg P \leqslant n$. Then for each $x \in D(\xi, p^{-l})$*

$$|P(x)|_p < (3k(n+1))^n N^{-v}.$$

This can be proved by modifying Lemma 2.10.

LEMMA 6.13. *Let $\delta > 0$, $\eta > 0$ be real numbers, $n \geqslant 2$ a natural number and $N = N(\delta, n)$ a sufficiently large real number. Further let $P, Q \in \mathbb{Z}[x]$ be two relatively prime polynomials of degree at most n and*

$$\max\{h(P), h(Q)\} \leqslant N.$$

Let $D = D(\xi, p^{-t})$ be a disc with $\mu(D) = p^{-t}$ where t is defined by the inequalities $p^{-t} \leqslant N^{-\eta} < p^{-t+1}$. If there exists a $\tau > 0$ such that for all $x \in D$

$$\max\{|P(x)|_p, |Q(x)|_p\} < N^{-\tau},$$

then $\tau + 2\max\{\tau - \eta, 0\} < 2n + \delta$.

Lemma 6.13 is proved in [42] (the real case is Lemma 2.9).

LEMMA 6.14. *Let $T_n(\eta)$ be the set of $\xi \in \mathbb{Q}_p$ for which the inequalties*

$$\left.\begin{array}{rcl} |P(\xi)|_p & < & h(P)^{-n-1}, \\ |P'(\xi)|_p & < & (\log h(P))^{-2-\eta} \end{array}\right\} \tag{6.12}$$

have infinitely many solutions in polynomials $P \in \mathbb{Z}[x]$ with $\deg P \leqslant n$. If $\eta > 0$ then $\mu(T_n(\eta)) = 0$.

The proof of this lemma is quite difficult although it has essentially the same logical structure as that of Lemma 2.7 and occasionally even has the same calculations. With the help of Lemma 6.11 we can pass from the system given by (6.12) to the system (6.9) with $\gamma = 2 + \eta$. Then for polynomials $P \in \mathfrak{I}_n$, define r_i and λ_i as in (6.5) and (6.6) and the classes of polynomials, $\mathfrak{I}_n(\mathbf{r})$ and $\mathfrak{I}_n(N, \mathbf{r})$. The magnitude of the components of the vector $\mathbf{r}$ is very important. If

$$r_2 T^{-1} + \lambda_1 > n, \tag{6.13}$$

then the proof of Lemma 6.14 follows from Sprindžuk's theorem [208, Part 2, Chapter 2, §10]. Next we consider the case

$$n - 1 + 2n\varepsilon_1 < r_2 T^{-1} + \lambda_1 \leqslant n. \tag{6.14}$$

First we will sketch why there can be only one polynomial belonging to a disc. It follows from (6.14), Lemmas 6.7, 6.8 and 6.9, that

$$|R(P, Q)|_p \leqslant N^{-2n-\varepsilon},$$

where $R(P,Q)$ is the resultant of the two polynomials P, Q which lie in the set

$$\bigcup_{N=2^t}^{2^{t+1}} \mathfrak{I}_n(N,\mathbf{r})$$

and which are small in a disc of fixed radius. But $|R(P,Q)|_p \gg N^{-2n}$. This contradiction shows that the disc has at most one polynomial belonging to it. Therefore we obtain an upper bound for the number of polynomials satisfying (6.14).

For the last two ranges

$$2-\varepsilon/2 \leqslant r_2T^{-1}+\lambda_1 \leqslant n-1+2n\varepsilon_1 \quad \text{and} \quad r_2T^{-1}+\lambda_1 < 2-\varepsilon/2,$$

the proof is analogous to Propositions 2.13 and 2.14. Combining these results we get the following.

LEMMA 6.15. *If $v > n+1$ then the Hausdorff dimension of $\widetilde{\mathcal{M}_v}$ and $\widetilde{\mathfrak{M}}_v$ is at most $(n+1)/v$.*

6.3.3. Lower bounds for Hausdorff dimension. The lower bound is less difficult than the upper and is obtained through approximation by algebraic numbers. For the rest of this chapter only let $\mathcal{K}_w$ be the set of $\xi \in \mathbb{Z}_p$ such that the inequality

$$|\xi-\alpha|_p < h(\alpha)^{-w}$$

has an infinite number of solutions in p-adic algebraic numbers α, with $\deg\alpha \leqslant n$; and let let $\omega(\xi)$ be the supremum of numbers w for which $\xi \in \mathcal{K}_w$. Again for the rest of this chapter only we write

$$\mathfrak{K}_v = \{\xi \in \mathbb{Q}_p \colon \omega(\xi) \leqslant v\} \quad \text{and} \quad \mathfrak{K}'_v = \{\xi \in \mathbb{Q}_p \colon \omega(\xi) = v\}.$$

The inequality $\dim \mathfrak{K}_v \geqslant (n+1)/v$ in [44] is the p-adic counterpart of the result in [13]. (The inequality $\dim \mathfrak{K}_v \leqslant (n+1)/v$ is not difficult and the proof is similar to those in §4.1.2 of Chapter 4.)

We will now give a brief exposition of [44] and use regular systems in $\mathbb{Q}_p$ to show that $\dim \widetilde{\mathcal{M}_v} \geqslant (n+1)/v$. Only those polynomials which obey the natural restrictions (6.4) need be considered.

6.3.4. Regular systems in $\mathbb{Q}_p$. A countable set Γ of p-adic numbers $\xi \in \mathbb{Q}_p$ together with a positive function G defined on Γ will be called a regular system (Γ, G) if for every disc $D = D(\xi, r)$ there is a positive number $N(D)$ such that for all $N \geqslant N(D)$ there are elements $\gamma_1, \dots, \gamma_t$ of Γ such that for each j, k with $1 \leqslant j$, $k \leqslant t$ and $j \neq k$, we have

$$\gamma_i \in D(\xi, r), \quad G(\gamma_i) \leqslant N, \quad |\gamma_j-\gamma_k|_p \geqslant N^{-1}, \quad t > c_1 rN,$$

where $c_1 = c_1(\Gamma, G)$ is a constant. It is clear that the same property then holds for any finite union of discs.

For any regular system (Γ, G) and any function $\psi\colon \mathbb{R}^+ \to \mathbb{R}^+$, we denote by $(\Gamma, G; \psi)$ the set of all p-adic numbers ξ for which there exist infinitely many γ

of Γ such that $|\xi - \gamma|_p < \psi(G(\gamma))$. Further, for any set Ω of p-adic numbers and any positive function g defined for $x > 0$ we shall write $\Omega \prec g$ if, for every positive λ and δ, Ω is covered by some countable set $\mathcal{D}_\delta(\lambda g)$ of discs, $D_1(\xi_1, \lambda_1)$, $D_2(\xi_2, \lambda_2), \dots$, with $\lambda_i \leqslant \lambda$ and

$$\sum_{i=1}^{\infty} g(\lambda_i) < \delta.$$

Clearly, if for sets Ω_1, Ω_2, we have $\Omega_1 \prec g$ and $\Omega_2 \prec g$, then $\Omega_1 \cup \Omega_2 \prec g$. Furthermore we have that $\Omega \prec g$ for every countable set Ω provided that g tends to zero as x does. Finally, if Ω has Hausdorff dimension d then $S \prec x^\rho$ for $\rho > d$ and $\Omega \not\prec x^\rho$ for $\rho < d$ since if S can be covered by a system $\mathcal{D}(\lambda, \rho) = \mathcal{D}_\infty(\lambda, x^\rho)$ for every $\rho > d$, $\lambda > 0$ then it can also be covered by a system $\mathcal{D}_\delta(\lambda, x^\rho)$ for every $\rho > d$, $\lambda > 0$, $\delta > 0$.

LEMMA 6.16. *Let $\psi\colon \mathbb{R}^+ \to \mathbb{R}^+$ be a decreasing function satisfying $x\psi(x) \leqslant 1/2$ for large x; and let $g\colon \mathbb{R}^+ \to \mathbb{R}^+$ be a decreasing function such that $x/g(x)$ increases and x and $x/g(x)$ tend to zero together. Assume also that $xg(1/2\psi(x))$ tends to infinity as x tends to infinity. Then for any regular system (Γ, G), we have that $(\Gamma, G; \psi) \not\prec g$. In fact, for any regular system (Γ_i, G_i), $i = 1, 2, \dots$ we have that*

$$\bigcap_{i=1}^{\infty} (\Gamma_i, G_i; \psi) \not\prec g.$$

In particular, the functions $\psi(x) = x^{-\sigma}$, $g(x) = x^s$, where $0 < s < \sigma^{-1} < 1$, satisfy the conditions of the lemma and we conclude therefore that the set of p-adic numbers ξ for which there exist infinitely many $\gamma \in \Gamma$ with

$$|\xi - \gamma|_p < (G(\gamma))^{-\sigma}$$

has dimension at least $1/\sigma$; this holds also for $\sigma = 1$ as then the dimension is clearly at least $1/\sigma'$ for every $\sigma' > 1$. Lemma 6.16 is a p-adic analogue of [13] and is proved in [44]. Note that $\Omega \not\prec x^s$ implies that for any λ-disc cover of Ω, the sum $\sum_{i=1}^{\infty} g(\lambda_i) \geqslant \delta$ and so the $\mathcal{H}^s(S) > 0$.

The lower bound for $\dim \mathfrak{K}_v$ depends on the set of algebraic p-adic numbers forming a regular system. In order to establish this we need the following two lemmas. The first is Hensel's lemma [61].

LEMMA 6.17. *Let P be a polynomial with coefficients in $\mathbb{Z}_p$, let $\xi = \xi_0 \in \mathbb{Z}_p$ and $|P(\xi)|_p < |P'(\xi)|_p^2$. Then as $n \to \infty$ the sequence*

$$\xi_{n+1} = \xi_n - \frac{P(\xi_n)}{P'(\xi_n)}$$

tends to some root $\beta \in \mathbb{Q}_p$ of the polynomial P and $|\beta - \xi|_p \leqslant |P(\xi)|_p / |P'(\xi)|_p^2 < 1$.

LEMMA 6.18. *Let D be a disc in $\mathbb{Q}_p$. Denote by $R_D(n, N, \delta)$ the set of $\xi \in D$, for which there are algebraic numbers $\alpha \in \mathbb{Q}_p$ with* $\deg \alpha \leqslant n$, $h(\alpha) \leqslant N$ *and*

$$|\xi - \alpha|_p < N^{-n-1}(\log N)^{4+\delta}.$$

Then for any $\delta > 0$, $\mu(R_D(n, N, \delta)) \to \mu(D)$ as $N \to \infty$.

PROOF. Let ξ be any transcendental number $\xi \in D \backslash R_D(n, N, \delta)$. There exists by Dirichlet's pigeonhole principle a polynomial $P(x) = a_n x^n + \cdots + a_1 x + a_0$ with integer coefficients and degree $\leqslant n$ such that

$$\left.\begin{array}{rcl} |P(\xi)|_p & \ll & N^{-n-1}, \\ |a_j| & \leqslant & N, \quad j = 0, 1, \ldots, n, \end{array}\right\} \tag{6.15}$$

where the implied constant depends only on n, δ and D. The height $h(P)$ of P is clearly $O(N)$, and with a suitable choice of constants, we can ensure that in fact $h(P) \leqslant N$.

We now distinguish between three cases. Assume first that $h(P) < \log N$. Since by Lemma 6.9, the root α of P nearest to ξ satisfies

$$|\xi - \alpha|_p \ll |P(\xi)|_p^{1/n} < N^{-1-1/n}$$

and as the number of polynomials P with $h(P) < \log N$ is $O(\log N)^{n+1}$, we see that ξ belongs to a set of measure $O\left(N^{-1-1/n}(\log N)^{2n}\right)$, which tends to zero as N tends to infinity.

Next assume that $h(P) \gg \log N$ and $|P'(\xi)|_p < (\log N)^{-2-\delta/2}$. Then we have the system of inequalities

$$\left.\begin{array}{rcl} |P(\xi)|_p & < & N^{-n-1}, \\ |P'(\xi)|_p & < & (\log N)^{-2-\delta/2}. \end{array}\right\} \tag{6.16}$$

Let $S_D(n, N, \delta)$ be the set of all ξ in D for which there exists a polynomial P with degree at most n, with integer coefficients and height at most N such that (6.16) is true. Then Lemma 6.14 shows that the Haar measure of

$$\bigcap_{1 \leqslant j \leqslant \infty} \bigcup_{N \geqslant j} S_D(n, N, \delta)$$

is zero and hence $\mu\left(\bigcup_{N \geqslant j} S_D(n, N, \delta)\right) \to 0$ as $j \to \infty$.

Finally assume that $|P'(\xi)|_p \geqslant (\log N)^{-2-\delta/2}$. From this inequality, (6.15) and Lemma 6.17 we deduce that there exists a root $\alpha \in \mathbb{Q}_p$ of a polynomial P such that

$$|\xi - \alpha|_p \ll \frac{|P(\xi)|_p}{|P'(\xi)|_p^2} \leqslant N^{-n-1}(\log N)^{4+\delta}.$$

Hence if N is sufficiently large, ξ lies in $R_D(n, N, \delta)$ which contradicts the initial assumption. This proves Lemma 6.18 □

Lemma 6.18 shows that the set Γ of all p-adic algebraic numbers $\alpha \in \mathbb{Q}_p$ with degree at most n together with the function $G(\alpha) = h(\alpha)^{n+1}(\log h(\alpha))^{-5}$ is a regular system. We have $R_D(n, N, \delta) \geqslant \mu(D)/2$ for all sufficiently large N. If we denote by $\{\gamma_1, \dots, \gamma_t\}$ a maximal subset of elements of Γ with $h(\gamma_i) \leqslant N$ and $|\gamma_k - \gamma_j|_p \geqslant N_1^{-1}$ for all j, k $(j \neq k)$, where $N_1 = N^{n+1}(\log N)^{-5}$, so that any γ of Γ with $h(\gamma) \leqslant N$ has distance $|\gamma - \gamma_i|_p \leqslant N_1^{-1}$ from one of the numbers $\gamma_1, \dots, \gamma_t$ then the union of the discs $D(\gamma, N_1^{-1})$ taken over all such γ has measure at most tN_1^{-1}, thus $\mu(R_D(n, N, \delta)) \leqslant tN_1^{-1}$ and $t > \mu(D)N_1/2$. We have a regular system with $c_1 = 1/2$.

6.3.5. Hausdorff dimension. The correct lower bound in Theorem 6.19 is obtained using Lemma 6.16. The functions f, g given by $f(x) = x^{-v/(n+1)} \log x$ for $x > 1$ and $g(x) = x^{(n+1)/v}$ satisfy the conditions of Lemma 6.16 and therefore the set of all p-adic numbers ξ for which there exist infinitely many algebraic p-adic numbers α of degree at most n and such that $|\xi - \alpha|_p < f(G(\alpha))$ has Hausdorff dimension at least $(n + 1)/v$. However

$$f(G(\alpha)) \ll h(\alpha)^{-v} (\log h(\alpha))^{5v/(n+1)+1}$$

and it follows that $\dim \widetilde{\mathcal{K}}_v$ and $\dim \widetilde{\mathfrak{K}}_v$ are at least $(n+1)/v$. For each $\xi \in \widetilde{\mathfrak{K}}_v$, there exists an algebraic p-adic number α of degree at most n such that $|\xi - \alpha| < h(\alpha)^{-v}$. Hence the minimal polynomial P for α satisfies $|P(\xi)|_p \leqslant |\xi - \alpha|_p$. Thus $\widetilde{\mathfrak{K}}_v \subset \widetilde{\mathscr{M}_v}$ and it follows that $\dim \widetilde{\mathscr{M}_v} \geqslant \dim \widetilde{\mathfrak{K}}_v \geqslant (n+1)/v$. As in the real case, it is evident that $\widetilde{\mathscr{M}'_v} \subset \widetilde{\mathscr{M}_v}$ and $\widetilde{\mathfrak{K}}'_v \subset \widetilde{\mathfrak{K}}_v$. By the choice of the function f, the Hausdorff measure $\mathcal{H}^{(n+1)/v}(\widetilde{\mathcal{K}}_v)$ is positive and so by §3.5.6 and the above we have the following.

THEOREM 6.19. *If* $v > n + 1$ *then*

$$\dim \widetilde{\mathfrak{K}}_v = \dim \widetilde{\mathfrak{K}}'_v = \dim \widetilde{\mathscr{M}_v} = \dim \widetilde{\mathscr{M}'_v} = \frac{n+1}{v}.$$

It would be interesting to extend these results, for example to curves lying in $\mathbb{Q}_p^2$, and establish the p-adic analogue of R. C. Baker's theorem [16] for planar curves). Having shown that such p-adic curves were extremal [160], Melnichuk used regular systems in $\mathbb{Q}_p$ to obtain the lower bound for the Hausdorff dimension of the corresponding exceptional set [162].

6.4. Notes

§**6.3** There are few results in p-adic metric Diophantine approximation on manifolds. Sprindžuk proved the analogue of Mahler's conjecture in $\mathbb{Q}_p$ amd I. L. Morotskaya has considered the inequality $|P(t)|_p < \psi(h(P))^n$ [164]; Melnichuk proved that certain curves were extremal in $\mathbb{Q}_p^2$ and found a lower bound for the Hausdorff dimension of points in the curves with a given degree of approximation [162].

The inhomogeneous case of the p-adic Sprindžuk theorem is proved in [38].

CHAPTER 7

Applications

7.1. Introduction

Constructing rational approximants of numbers, such as π and the Metonic ratio, which occur in geometry and in the physical world, relies on Diophantine approximation of some sort. It is likely that Diophantine approximation was used to design sophisticated gearing 2500 years ago [185]. Other applications arise from the natural link between rational dependence (corresponding to a Diophantine equation) and the physical phenomenon of resonance and so between Diophantine approximation and proximity to resonance. This can give rise to the 'notorious problem of small denominators' [1] in which solutions contain denominators that can become arbitrarily small, thus making convergence problematic. Small denominators occur in the study of the stability of the solar system (the N-body problem) and in dynamical systems; the fundamental character of these problems has inspired much new mathematics (fuller accounts are in [6], [9], [167], [233], [234]). They also occur in related questions, such as averaging, and linearisation and normal forms.

The basic idea is to exclude sets of denominators which prevent convergence without significantly affecting the validity of the solution. Small denominators are related to very well approximable points, which lie in the complements of sets of points of certain Diophantine type; these complements include badly approximable numbers and their higher dimensional counterparts. The techniques developed in the metric theory of Diophantine approximation lend themselves to the study of the distribution of small denominators and in particular to the analysis of the Hausdorff dimension of the associated exceptional sets. Some examples of problems involving small denominators and the associated exceptional sets are discussed below.

Classical Diophantine approximation also offers a model for approximation in more general settings. The prime example of this is discrete groups acting on hyperbolic space, a smooth manifold with constant negative curvature, and a short account of approximation in hyperbolic space is given in §7.7. Note that hyperbolic space is not regarded as having been embedded in in Euclidean space [170]; the boundaries of the disc and half-plane models are taken as a setting for a generalisation of simultaneous Diophantine approximation in Euclidean space.

[1] The terminology 'small divisors' is also used but could be confusing in a number theory book.

7.2. Diophantine type and very well approximable numbers

The real number ξ is said to be of *Diophantine type* (K, σ) [7] if

$$\left|\xi - \frac{p}{q}\right| \geqslant \frac{K}{q^{\sigma+2}}$$

for all rationals p/q; the set of such is denoted $\mathcal{D}(K, \sigma)$. The union

$$\mathcal{D}(\sigma) = \bigcup_{K>0} \mathcal{D}(K, \sigma)$$

consists of numbers of Diophantine type (K, σ) for some $K > 0$ and the union

$$\mathcal{D} = \bigcup_{\sigma>0} \mathcal{D}(\sigma) = \bigcup_{K,\sigma>0} \mathcal{D}(K, \sigma)$$

is the set of numbers of Diophantine type for some (K, σ). The set $\mathcal{D}(\sigma)$ and its higher dimensional analogues are complementary to the sets of points approximable to exponent $\sigma + 2$ and in a sense to sets of very well approximable points. Points in $\mathcal{D}(\sigma)$ have desirable approximation properties for certain applications. Note that $\mathcal{D}(0) = \mathfrak{B}$, the set of badly approximable numbers.

Given $\sigma > 0$, almost all real numbers are of Diophantine type (K, σ) for some positive K, *i.e.*, the set

$$\mathcal{D}(\sigma) = \bigcup_{K>0} \{\xi \in \mathbb{R}\colon |\xi - p/q| \geqslant Kq^{-2-\sigma} \text{ for each } p/q \in \mathbb{Q}\}$$

is of full measure, since the complementary set

$$E_\sigma = \bigcap_{K>0} \{\xi \in \mathbb{R}\colon |\xi - p/q| < Kq^{-2-\sigma} \text{ for some } p/q \in \mathbb{Q}\}$$

is null and $\dim E_\sigma = 2/(\sigma + 2)$. This is a consequence of the following inclusion argument. Given $\varepsilon > 0$,

$$\mathscr{K}_{\sigma+1+\varepsilon} \subset E_\sigma \subset \mathscr{K}_{\sigma+1},$$

where $\mathscr{K}_v$ is the set of v-approximable numbers. For let $\xi \in \mathbb{R} \setminus \mathscr{K}_{\sigma+1}$, so that for all but finitely many positive integers q, say $q_1, \ldots, q_r$,

$$\|q\xi\| = |\langle q\xi\rangle| < q^{-\sigma-1}.$$

Then $K = K(\xi) = \min\{\|q_j\xi\| q_j^{\sigma+1}\colon j = 1, \ldots, r\} > 0$ since $\langle q\xi\rangle = 0$ implies $\langle kq\xi\rangle = 0$ for each $k \in \mathbb{Z}$. Thus $\|q\xi\| \geqslant K\, q^{1+\sigma}$ for all $q \in \mathbb{N}$ and so $\xi \notin E_\sigma$.

To establish the other inclusion, let $\xi \in \mathscr{K}_{\sigma+1+\varepsilon}$, so that

$$|\langle q\xi\rangle| < q^{-\sigma-1-\varepsilon}$$

for infinitely many $q \in \mathbb{N}$. Given any $K > 0$, choose a q_0 from these q sufficiently large so that $q_0^{-\varepsilon} \leqslant K$. Then there exists a $p_0 \in \mathbb{Z}$ such that

$$\|q_0\xi\| = |q_0\xi - p_0| < K\, q_0^{-\sigma-1}$$

and $\xi \in E_\sigma$. It follows from these inclusions, Theorem 3.5 and the Jarník-Besicovitch theorem in Chapter 3 that when $\sigma \geqslant 1$,

$$\frac{2}{\sigma+2+\varepsilon} \leqslant \dim E_\sigma \leqslant \frac{2}{\sigma+2},$$

whence $\dim E_\sigma = 2/(\sigma+1)$ as ε is arbitrary. Let $E = \bigcap_{\sigma>0} E_\sigma = \lim_{\sigma\to\infty} E_\sigma$ since E_σ decreases as σ increases. Then

$$\dim E = \lim_{\sigma\to\infty} \frac{2}{\sigma+1} = 0, \tag{7.1}$$

and the complement of $\mathcal{D}$ has Hausdorff dimension 0.

7.3. A wave equation

As a relatively accessible illustration of the way small denominators arise and the way Diophantine analysis is used to discuss them, we will start with a partial differential equation. Let α, β be two positive numbers and let $g\colon \mathbb{R}^2 \to \mathbb{C}$ be a C^∞ doubly periodic function with periods α, β and Fourier series expansion

$$g(x,t) = \sum_{j,k\in\mathbb{Z}} g_{jk} e^{2\pi i(jx/\alpha + kt/\beta)}. \tag{7.2}$$

As can be verified by repeated integration by parts, the coefficients g_{jk} decay very rapidly: when j, k are non-zero and for any N

$$|g_{j,k}| \ll (|j||k|)^{-N}. \tag{7.3}$$

When either j or k vanishes, $|g_{jk}|$ is $O(|k|^{-N})$ or $O(|j|^{-N})$ respectively.

The inhomogeneous partial differential equation

$$\frac{\partial^2 f}{\partial x^2} - \frac{\partial^2 f}{\partial t^2} = g \tag{7.4}$$

describes the transmission of a wave f from a source g. It can be solved formally by substituting the Fourier series

$$f(x,t) = \sum_{j,k\in\mathbb{Z}} f_{jk} e^{2\pi i(jx/\alpha + kt/\beta)}$$

for f into (7.2) and comparing coefficients. This gives

$$(-4\pi^2) \sum_{j,k\in\mathbb{Z}} \left(\frac{j^2}{\alpha^2} - \frac{k^2}{\beta^2}\right) f_{jk} e^{2\pi i(jx/\alpha + kt/\beta)} = \sum_{j,k\in\mathbb{Z}} g_{jk} e^{2\pi i(jx/\alpha + kt/\beta)},$$

whence

$$f_{jk} = \frac{\alpha^2 g_{jk}}{4\pi^2(k^2 - \gamma^2 j^2)},$$

where $\gamma = \alpha/\beta$, so that we must have $g_{00} = 0$. The series

$$f(x,t) = \frac{\alpha^2}{4\pi^2} \sum_{j,k\in\mathbb{Z}} \frac{g_{jk}}{k^2 - \gamma^2 j^2} e^{2\pi i(jx/\alpha + kt/\beta)} \tag{7.5}$$

solves (7.4) and satisfies

$$|f(x,t)| \ll \sum_{j\neq 0} |f_{0j}| + \sum_{k\neq 0} |f_{k0}| + \sum_{j,k\neq 0} |f_{jk}|.$$

It is readily verified that the series for f and its derivatives $\partial^{r+s} f/\partial x^r \partial t^s$ converge providing

$$\sum_{j\neq 0} \frac{|j|^r |g_{j0}|}{(\gamma j)^2} + \sum_{k\neq 0} \frac{|k|^s |g_{0k}|}{k^2} + \sum_{j\neq 0}\sum_{k\neq 0} \frac{|j|^r |k|^s |g_{jk}|}{|(k-\gamma j)(k+\gamma j)|} < \infty.$$

But $|g_{j0}| \ll |j|^{-N}$ for any positive N and so choosing $N > r-1$, we get that

$$\sum_{j\neq 0} \frac{|j|^r |g_{j0}|}{(\gamma j)^2} \ll \sum_{j\neq 0} |j|^{r-2-N} < \infty$$

and similarly $\sum_{k\neq 0} |k|^{s-2} |g_{0k}|$ converges for $N > s-1$.

The small denominator problem arises with the double sum (over $j, k \neq 0$), since

$$\sum_{j,k\neq 0} |f_{jk}| \ll \sum_{j,k\neq 0} \frac{|j|^r |k|^s |g_{jk}|}{|(k-\gamma j)(k+\gamma j)|} \ll \sum_{k=1}^{\infty} k^{s-N} \sum_{j=1}^{\infty} \frac{|j|^{r-N}}{|(k-\gamma j)(k+\gamma j)|}$$

by (7.3). The sum over $j, k \geqslant 1$ converges providing in addition that

$$|(k-\gamma j)| \geqslant \|\gamma j\| \gg |j|^{-v}$$

for some $v > 0$. Hence when $\gamma = \alpha/\beta$ is of Diophantine type $(K, v+2)$, f and its partial derivatives up to order r, s are well defined since we can choose $N > \max\{v+s+2, r+2\}$. Thus when g is C^∞ and α/β is of Diophantine type $(K, v+2)$ for any positive K, v, the solution f is also C^∞. It follows that the partial differential equation (7.4) has a C^∞ solution f given by (7.5) when g is C^∞ with no constant term and when γ is *not* a Liouville number. For a more complete account see [176].

The set of Liouville numbers is of zero Lebesgue measure and zero Hausdorff dimension as well (see §3.5.3). Thus for almost all α/β, the partial differential equation (7.4) has a C^∞ solution f given by (7.5) when g is C^∞ with no constant term; and the Hausdorff dimension of the exceptional set where the solution does not necessarily exist is 0.

7.4. The rotation number

As another relatively accessible illustration of the way small denominators arise in normal form questions, we will discuss the rotation number of an orientation preserving circle diffeomorphism. More details are given in [8], [175], [233] and there is a simplified account in [100]. First we need to set some notation. Let $f: S^1 \to S^1$ be a continuous circle map. Then f has a continuous lift $F: \mathbb{R} \to \mathbb{R}$ which satisfies $e \circ F = f \circ e$, where $e: \mathbb{R} \to S^1$ is the exponential map given by

$$e(t) = e^{2\pi i t}.$$

Lifts are unique up to translation by an integer. When f is an orientation preserving circle homeomorphism, F is strictly increasing and $F - 1$ (given by $F(t) - t$) is 1-periodic. Moreover the limit

$$\rho(F) = \lim_{N\to\infty} \frac{F^N(t)}{N}$$

exists for each $t \in \mathbb{R}$ and is independent of t. Here F^N is the N-fold iteration $F \circ \cdots \circ F$. The rotation number of a rotation r_ρ by an angle $2\pi\rho$ given by

$$r_\rho(z) = z\, e(\rho) = ze^{2\pi i\rho}$$

is ρ (we follow [175, Chapter 1] here; the notation in [8] differs slightly).

The rotation number $\rho(F)$ for F is rational if and only if f has a periodic point. Two rotation numbers $\rho(F)$ and $\rho(F')$ for lifts F, F' for f differ by an integer and so the rotation number $\rho(f)$ for f can defined as the fractional part $\{\rho(F)\}$ of $\rho(F)$. Thus $\rho(f)$ is irrational if and only if f has no periodic points. If $\rho(f)$ is irrational then for $z \in S^1$, the closure of the orbit

$$\omega(z) = \{f^n(z) \colon n \in \mathbb{N}\}$$

does not dependent on z and either is perfect and nowhere dense or is S^1. In the latter case f is called transitive and is topologically conjugate to the rotation r_ρ, where $\rho = \rho(f)$, *i.e.*, there exists an orientation preserving homeomorphism $\varphi\colon S^1 \to S^1$ such that

$$f = \varphi^{-1} \circ r_{\rho(f)} \circ \varphi \quad \text{or} \quad \varphi \circ f = r_{\rho(f)} \circ \varphi, \tag{7.6}$$

written $f \sim r_{\rho(f)}$. A. Denjoy showed that when f is C^2, the irrationality of $\rho(f)$ implies that $f \sim r_{\rho(f)}$, so that f is topologically conjugate to a rotation. This is best possible as there are counterexamples when f is $C^{2-\varepsilon}$ [233].

Two kinds of Diophantine approximation enter the picture when the differentiability of the conjugacy function φ is considered. When φ is C^r, where r is finite or infinite (but φ is not analytic), the existence or otherwise of a smooth conjugacy is determined by the Diophantine type of ρ. If f is C^r and $\rho(f)$ is of Diophantine type (*i.e.*, $\rho \in \mathcal{D}$), then f is smoothly conjugate to $r_{\rho(f)}$. The precise details of the differentiability conditions are rather complicated; further details and references are in [233].

When φ is analytic and ρ is a Bryuno number, *i.e.*, $\rho \in \mathbb{R} \setminus \mathbb{Q}$ and the denominator q_n of the n-th convergent of $\xi \in \mathbb{R} \setminus \mathbb{Q}$ satisfies

$$\sum_{n=1}^{\infty} \frac{\log q_{n+1}}{q_n} < \infty,$$

then any analytic f with rotation number ρ is analytically conjugate to r_ρ. This condition is also optimal since if ρ is not a Bryuno number, there is a Blaschke product with rotation number ρ not analytically conjugate to r_ρ [233]. The set of Bryuno numbers includes $\mathcal{D}$ since if $\xi \in \mathcal{D}(K, \sigma)$ then $q_{n+1} = O(q_n^\sigma)$ and $\sum_n((\log q_{n+1})/q_n) < \infty$.

To illustrate the ideas, we shall follow [8] and use Diophantine type conditions when considering analytic conjugation when f is close to a rotation; this can be regarded as a stability result. Let ρ be of Diophantine type (K, σ). Suppose the diffeomorphism f is analytic with (analytic) lift F given by

$$F(x) = x + \rho + a(x)$$

where a is a 1-periodic analytic function, real on the real axis and bounded on the strip $\{z \in \mathbb{C}\colon |\operatorname{Im} z| < \varepsilon\}$ for some $\varepsilon > 0$. Then there exists a positive $\delta = \delta(K, \sigma, \varepsilon)$ such that if

$$\sup\{|a(z)|\colon |\operatorname{Im} z| < \varepsilon\} < \delta, \tag{7.7}$$

then f is analytically conjugate to a rotation by ρ, *i.e.*, there exists an analytic homeomorphism φ such that $f \circ \varphi(z) = \varphi \circ r_\rho(z)$. To see this, consider the lifts F, Φ and R_ρ of f, φ and r_ρ and the lift

$$F \circ \Phi(t) = \Phi \circ R_\rho(t) \tag{7.8}$$

of equation (7.6), where R_ρ is the translation given by $R_\rho(t) = t + \rho$ and where Φ is assumed to be close to the identity, *i.e.*, $\Phi(t) = t + h(t)$, where $h(t)$ is small. Then (7.8) reduces to the functional equation

$$t + h(t) + \rho + a(t + h(t)) = t + \rho + h(t + \rho)$$

i.e., to

$$h(t + \rho) - h(t) = a(t + h(t)). \tag{7.9}$$

One then seeks a solution to this equation by successive approximations. Since a and h are small, consider the first order approximation

$$h^0(t + \rho) - h^0(t) = \tilde{a}(t) = a(t) - a_0, \tag{7.10}$$

where $a_0 = \int_0^1 a(t)dt$, the mean of $a(t)$. Since a and h are periodic, they have Fourier series expansions

$$a(t) = \sum_{k \in \mathbb{Z}} a_k e^{2\pi i k t} \quad \text{and } h^0(t) = \sum_{k \in \mathbb{Z}} h^0_k e^{2\pi i k t}.$$

On substituting in (7.10) and comparing the coefficients of $e^{2\pi i k t}$, the coefficients of the linear approximation h^0 to the unknown function h can be determined in terms of those of a: $h^0_0 = 0$ and for $k \neq 0$,

$$h^0_k = \frac{a_k}{e^{2\pi i k \rho} - 1}.$$

It is evident that for h^0 to exist, the denominator and numerator of h^0_k must vanish together (this why the function $\tilde{a}(t) = a(t) - a_0$ is introduced). Since a is analytic, its Fourier coefficients decay rapidly and $|a_k| < \delta\, e^{-|k|\varepsilon}$ when a satisfies (7.7) [8, §12D]. Now given any real θ, it is readily verified (since $2\theta/\pi \leqslant \sin\theta \leqslant \theta$ when

$0 \leqslant \theta \leqslant \pi/2$) that $|e^{i\theta} - 1| \geqslant 2|\langle\theta\rangle|/\pi$, whence (since ρ is of Diophantine type (K, σ)) for $k \neq 0$,

$$|e(k\rho) - 1| = |e^{2\pi i k\rho} - 1| \geqslant 2|\langle k\rho\rangle|/\pi \geqslant 2K|k|^{-\sigma}/\pi.$$

Thus the Fourier series for h^0 converges for such ρ when $\sigma > 0$. Successive approximations to the function Φ are then constructed using a version of Newton's tangent method. First the function $\Phi_0 = 1 + h^0$ is invertible. Let

$$A_1 = \Phi_0^{-1} \circ R_\rho \circ \Phi_0$$

and let a^1 be defined by $A_1(t) = t + \rho + a^1(t)$, solve the modified equation (7.10)

$$h^1(t + \rho) - h^1(t) = \tilde{a}^1(t) = a^1(t) - a_0^1$$

and so on. The sequence of functions Φ_k so constructed converges to an analytic function Φ satisfying (7.9) when ρ is of type (K, σ) for some K, σ and a is sufficiently small (see [8, §12]). Hence the desired conjugacy exists, *providing* ρ is of type (K, σ), *i.e.*, providing $\rho \in \mathcal{D}$.

The corresponding exceptional complementary set $E = \mathbb{R} \setminus \mathcal{D}$ is given by

$$E = \bigcap_{\sigma > 0} E_\sigma = \bigcap_{\sigma \in \mathbb{Q}^+} E_\sigma = \lim_{\sigma \to \infty} E_\sigma$$

and has zero Hausdorff dimension by (7.1). Since the complement of $\mathcal{B}$ is contained in E, the complement of the set $\mathcal{B}$ is also of Hausdorff dimension 0. Thus when f is analytic, f is analytically conjugate to a rotation unless the rotation number of f lies in a set of Hausdorff dimension 0.

7.5. Dynamical systems

It will be helpful in discussing other topics to say a little about flows.

7.5.1. Quasi-periodic flows on the torus. For each vector $\omega = (\omega_1, \dots, \omega_n)$ in the the n-dimensional torus $\mathbb{T}^n = \mathbb{S}^1 \times \cdots \times \mathbb{S}^1$, the map $\varphi_\omega \colon \mathbb{R} \to \mathbb{T}^n$ given by

$$\varphi_\omega(t) = \varphi_\omega(0) + t\omega \tag{7.11}$$

is called a *quasi-periodic* flow on the torus. The coordinates of the point ω are called frequencies. When the frequencies are all rational, the flow is a closed path on the torus and returns to its starting point after a time equal to the lowest common multiple of the denominators of the frequencies. Thus if $\langle \mathbf{q} \cdot \omega \rangle = 0$, the resonant set $R_{\mathbf{q}} = \{x \in \mathbb{T}^n \colon \langle \mathbf{q} \cdot x \rangle = 0\}$ coincides with $\varphi(\mathbb{R})$. If the frequencies are not all rational, then by Kronecker's theorem [114], the flow winds round the torus, densely filling a subspace of dimension given by the number of rationally independent frequencies. Thus when the frequencies are independent over the rationals, the closure of $\varphi_\omega(\mathbb{R})$ is the torus $\mathbb{T}^n$. (Note the similarity with conjugacy above.)

7.5.2. Kolmogorov-Arnol'd-Moser theory. One of the oldest problems in mechanics has been to understand the motions of N bodies subject to Newtonian attraction. The solution is well known for $N = 2$ and periodic solutions exist, with the bodies moving in elliptic orbits about their centre of mass. In the absence of any effects such as friction, the solution persists for all time. For $N \geqslant 3$, the position is extraordinarily complicated and far from understood. In celestial mechanics, a particularly interesting case is that of a solar system in which one of the bodies is a sun with mass m_N very much larger than the other masses $m_1, \ldots, m_{N-1}$ of the planets. If as a first approximation, the centre of mass of the system is assumed to coincide with that of the sun and if the gravitational interactions between the planets and other effects are neglected, then the system decouples into $N - 1$ two-body problems, in which each planet describes an elliptical orbit around the sun, with period T_j say and frequency $\omega_j = 2\pi/T_j$, $j = 1, \ldots, N-1$. Such a solution can be described by the quasi-periodic flow φ_ω on the torus $\mathbb{T}^n$, where $n = N - 1$, given by $\dot{\varphi}_\omega = \omega = (\omega_1, \ldots, \omega_n)$ or by (7.11), and will persist for all time.

The gravitational interactions between the planets can be taken into account by considering a perturbation and it is of interest to know whether the motion of the planets remains quasi-periodic. To study this question, H. Poincaré developed a wealth of new and important ideas and results [182]. His theorem on the integrals in the three body problem suggested that quasi-periodic solutions could not exist. K. Weierstrass [167, Chapter 1, §2] constructed a formal series solution but was unable to establish convergence because the denominators contained factors of the form $\mathbf{q} \cdot \omega$ which can become very small for certain integer vectors $\mathbf{q}$. The problem was resolved in 1962 when V. I. Arnol'd and A. N. Kolmogorov not only established the existence of such solutions for a perturbed analytic Hamiltonian system but also showed that they were relatively abundant in the sense that the set of perturbed solutions that are quasi-periodic forms a complicated Cantor type set of positive Lebesgue measure [167, p. 8]. It follows that for planets with mass very much less than the sun and for the majority of initial conditions in which the orbits are close to co-planar circles, it can be shown that the planets will never collide, escape or fall into the sun.

The approach of Kolmogorov and Arnol'd was to estimate the set of initial values of the system which generate stable motion. First the frequencies of the perturbed Hamiltonian are chosen to satisfy a Diophantine condition which holds for almost all frequencies ω and also guarantees convergence in the analysis (this involves an infinite dimensional extension of Newton's tangent method). Thus this condition guarantees stable motion. The frequencies are kept equal to those of the unperturbed system and new initial conditions are sought.

A little background is required (more details and references are in [96]). The equations which describe the behaviour of general non-dissipative mechanical systems with n degrees of freedom can be transformed into a Hamiltonian system of partial differential equations

$$\dot{x}_k = \frac{\partial H}{\partial y_k}, \quad \dot{y}_k = -\frac{\partial H}{\partial x_k}, \quad k = 1, \ldots, n,$$

where $H\colon \mathbb{R}^n \times \mathbb{R}^n \to \mathbb{R}$. The variables x are generalised angles and the y generalised momenta. The Hamiltonian system is called *integrable* if there is a *canonical* transformation $W\colon \mathbb{R}^n \times \mathbb{T}^n \to \mathbb{R}^n \times \mathbb{R}^n$ which preserves the symplectic structure and relates the coordinates $(x, y) = W(I, \Theta)$ of a point in phase space to *action-angle* coordinates (I, Θ). In the action-angle coordinates W is periodic in the angle variable Θ, $H \circ W$ is independent of Θ and Hamilton's equations become

$$\dot{\Theta}_k = \frac{\partial H}{\partial I_k}, \quad \dot{I}_k = -\frac{\partial H}{\partial \Theta_k} = 0, \quad k = 1, \ldots, n \tag{7.12}$$

(we follow the usual practice in physics of writing the transformed Hamiltonian $H \circ W$ simply H). These equations are solved by

$$I_k = c_k, \quad \Theta_k = \omega_k t + \Theta_k(0), \quad k = 1, \ldots, n,$$

where $\omega_k = \partial H / \partial I_k(c)$ and $c = (c_1, \ldots, c_n)$ is a constant vector. This solution is a quasi-periodic flow on the torus $\mathbb{T}^n$. Each level set given by $I = c$ corresponds to an n-dimensional torus in phase space $\mathbb{R}^{2n}$. The n frequencies are the derivatives of the Hamiltonian H and so depend only on the action I. The quasi-periodic flow remains on the torus which is thus invariant under the flow.

The planetary interactions are represented by a small perturbation of the original Hamiltonian. We work in an open bounded subset Y (corresponding to momenta) of $\mathbb{R}^n$ and consider the perturbation $H_\varepsilon\colon Y \times \mathbb{T}^n \to \mathbb{R}$ given by

$$H_\varepsilon(I, \Theta) = H_0(I) + \varepsilon \widetilde{H}(I, \Theta, \varepsilon),$$

where $H_0(I)$ is the original unperturbed analytic Hamiltonian and where $\widetilde{H}$ is analytic. The stability of the solar system then reduces to the question of whether the solutions of the perturbed Hamiltonian system still wind round a perturbed invariant torus.

Let $\mathcal{T}_\varepsilon$ be the set of invariant tori of the system (7.12) filled with quasi-periodic solutions and lying in $Y \times \mathbb{R}^n$. Diophantine approximation is again used to avoid the convergence problems in the study of $\mathcal{T}_\varepsilon$. It is assumed that ω satisfies a higher dimensional version of Diophantine type discussed in §7.2 above. To be precise it is assumed that there exist positive constants $K = K(\omega)$ and $v = v(\omega)$ such that for all non-zero $\mathbf{q} \in \mathbb{Z}^n$, the Diophantine condition

$$|\mathbf{q} \cdot \omega| \geqslant K |\mathbf{q}|_1^{-v} \tag{7.13}$$

holds. The exponent v is subject to two conflicting requirements. It should be large enough ($v > n$) to ensure that the Diophantine condition above is not too restrictive but small enough to ensure that the perturbation has physical significance and that the stability is robust.

The next step is to choose a vector $c \in \mathbb{R}^n$ satisfying $\partial H(c)/\partial I_j = \omega_j$, $j = 1, 2, \ldots, n$, where ω satisfies (7.13). This can be done when the Hamiltonian H is non-degenerate, *i.e.*, when the Hessian $\det(\partial^2 H/\partial I_j \partial I_k) \neq 0$. The condition (7.13) also serves to guarantee the convergence of an infinite dimensional version of Newton's tangent method and indeed turns out to be a natural requirement for the application of an infinite dimensional implicit function theorem [225]. Then it can be shown that as $\varepsilon \to 0$, $|(Y \times \mathbb{T}^n) \setminus \mathcal{T}_\varepsilon| < \varepsilon |Y \times \mathbb{T}^n|$, whence the Lebesgue measure of $\mathcal{T}_\varepsilon$ tends to full measure in $Y \times \mathbb{T}^n$ as $\varepsilon \to 0$ (see [55], [167]).

It can be seen that the majority of orbits are quasi-periodic and the result holds providing the frequencies are not near to resonance. However, the proof breaks down when the frequencies lie in the complementary exceptional set E_v say of frequencies which are close to resonance in the sense that given any $C > 0$, there exists a $\mathbf{q} \in \mathbb{Z}^n$ such that

$$|\mathbf{q} \cdot x| < C|\mathbf{q}|_1^{-v}.$$

This set is closely related to the set

$$\widehat{\mathscr{L}_v}(\mathbb{R}^n) = \{x \in \mathbb{R}^n \colon |\mathbf{q} \cdot x| < |\mathbf{q}|_1^{-v} \text{ for infinitely many } \mathbf{q} \in \mathbb{Z}^n\} \qquad (7.14)$$

and in fact given any $\varepsilon > 0$,

$$\widehat{\mathscr{L}}_{v+\varepsilon}(\mathbb{R}^n) \subset E_v \subset \widehat{\mathscr{L}_v}(\mathbb{R}^n),$$

as can be seen by an inclusion argument very similar to the one in the preceding section (see [91]). It follows that the two sets have the same Hausdorff dimension.

The transformation $T \colon (1/4, 1/2) \times [-1, 1]^{n-1} \to \mathbb{R}^n$ given by

$$T(x_1, \ldots, x_n) = (x_1, x_2/x_1, \ldots, x_n/x_1)$$

is bi-Lipschitz and given any $\varepsilon > 0$ it is readily verified that

$$T\left(\widetilde{\mathcal{L}}_v(\mathbb{R}^n) \cap \left((1/4, 1/2) \times [-1, 1]^{n-1}\right)\right) \supseteq (1/4, 1/2) \times \mathscr{L}_{v+\varepsilon}([-1, 1]^{n-1}),$$

where $\mathscr{L}_v([-1, 1]^{n-1})$ is the set of $x \in [-1, 1]^{n-1}$ such that the inequality

$$\|x \cdot \mathbf{q}\| < |\mathbf{q}|^{-v}$$

holds for infinitely many $\mathbf{q} \in \mathbb{Z}^{n-1}$ (see §1.3.2). It follows from Theorem 3.7 and §3.4 that for each $\varepsilon > 0$,

$$\dim \widehat{\mathscr{L}}_{v+\varepsilon}(\mathbb{R}^n) \geqslant 1 + \dim \mathscr{L}_{v+\varepsilon}([-1, 1]^{n-1}) = n + \frac{n}{v + \varepsilon + 1},$$

whence since $\varepsilon > 0$ is arbitrary, $\dim \widehat{\mathscr{L}_v} \geqslant n + n/(v + 1)$. The complementary inequality can be established directly using a standard covering argument [96] and it follows that

$$\dim E_v = \dim \widehat{\mathscr{L}_v}(\mathbb{R}^n) = n - 1 + \frac{n}{v + 1}.$$

Diophantine approximation of the kind

$$|\mathbf{q} \cdot x| < \psi(|\mathbf{q}^{(1)}|_1 + |\mathbf{q}^{(2)}|_1),$$

where $\mathbf{q} = (\mathbf{q}^{(1)}, \mathbf{q}^{(2)}) \in \mathbb{Z}^m \times \mathbb{Z}^k$ and $|\mathbf{q}^{(1)}|_1 \leqslant h$, on a manifold of the type $M = \{(u, \theta(u) \colon u \in U\}$ arises in connection with problems of small denominators in infinite dimensional Kolmogorov-Arnol'd-Moser theory [87]. The approximation falls between the independent ($h = 0$) and dependent cases ($h = \infty$). The Hausdorff dimension of the exceptional set

$$W_h(\psi) = \{x \in M \colon |\mathbf{q} \cdot x| < \psi(|\mathbf{q}|_1) \text{ for infinitely many } \mathbf{q} \in \mathbb{Z}^m \times \mathbb{Z}_h^k\},$$

where $\mathbb{Z}_h^k = \{\mathbf{q}^{(2)} \colon |\mathbf{q}^{(2)}| \leqslant h\}$, is obtained using ubiquity: when $\sum_{r=1}^{\infty} r^{m-1}\psi(r)$ converges, the Hausdorff dimension is $m - 1 + m/(\lambda + 1)$, where $\lambda \geqslant m - 1$ is the lower order of ψ, and does not depend on h.

7.5.3. Averaging in dynamical systems. Diophantine approximation also arises in the technique of 'averaging' in asymptotic methods of perturbation theory [8, Chapter 4]. The underlying idea is that a perturbed differential equation is replaced by a much simpler equation in which a spatial average is substituted for a time average. This substitution is not valid near resonances. For example consider the differential equation perturbed by a small rapid oscillation:

$$\dot{x} = \varepsilon f(x, y, \varepsilon), \quad \dot{y} = \omega(x) + \varepsilon g(x, y, \varepsilon) \tag{7.15}$$

where $x \in D$, a bounded domain in $\mathbb{R}^m$, $y \in \mathbb{T}^n = \mathbb{R}^n/\mathbb{Z}^n$, ε is a sufficiently small positive parameter, $\omega \colon D \to \mathbb{R}^n$ ($\mathbb{R}^n$ is the tangent space to $\mathbb{T}^n$) and the functions f, g, ω are continuously differentiable with respect to x, y and ε. The perturbed Hamiltonian systems discussed in the preceding section are an example of (7.15) with $m = n$ and small rapidly oscillating part $\varepsilon \widetilde{H}(I, \Theta, \varepsilon)$. These differential equations can be studied by replacing (7.15) by the simpler equation

$$\dot{u} = \varepsilon \int_{\mathbb{T}^n} f(x, y, \varepsilon)\, dy,$$

obtained by averaging over the angular variables y.

Let $\rho_0 > 0$ and let $G \subset D$ be a compact convex domain consisting of those initial conditions (u^0, y^0) for which $u(t)$ is defined for all $t \in [0, 1/\varepsilon]$ and gets no closer than $2\rho_0$ to the boundary of D. Solutions of

$$x(t) = x(t, x^0, y^0, \varepsilon), \quad u(t) = u(t, x^0, \varepsilon)$$

with identical initial conditions $(x^0, y^0) \in G \times \mathbb{T}^m$ ($y^0 = \omega$ in the preceding section) are considered. Define

$$h(x^0, y^0; \varepsilon) = \sup\{|x(t) - u(t)| \colon t \in [0, 1/\varepsilon]\}$$

and let

$$\mathscr{I}(\rho, \varepsilon) = \{(x^0, y) \in G \times \mathbb{T}^n \colon |h(x^0, y^0; \varepsilon)| > \rho \text{ for } t - t_0 \leqslant 1/\varepsilon\}.$$

A. Neishtadt [173] proved that when the frequencies $\omega_1(x), \ldots, \omega_n(x)$ are independent (*i.e.*, when the rank of the Jacobian $(\partial\omega_j/\partial x_k)$ is n), then for sufficiently small positive ε, ρ,

$$|\mathscr{I}(\rho,\varepsilon)| < K_0\,\varepsilon^{1/2}\,\rho^{-1} \tag{7.16}$$

and showed that this estimate was best possible. This result depends on estimating the measure of the set

$$V(\delta) = \left\{x \in D \colon |\mathbf{q}\cdot\omega(x)| \leqslant \delta\,|\mathbf{q}|^{-n} \text{ for some } \mathbf{q} \in \mathbb{Z}^n \setminus \{0\}\right\}$$

and it can be shown that

$$|V(\delta)| = O(\delta), \tag{7.17}$$

However, if $n > m$, *i.e.*, if the number n of angular variables exceeds m, then $\operatorname{rank}(\partial\,\omega_j/\partial x_k) \leqslant \min\{m,n\} < n$. Thus the frequencies $\omega(x)$ cannot be independent and (7.17) is not valid in general.

When $m < n$, the set D can be taken to be a parametrisation domain and $\omega \colon D \to \mathbb{R}^n$ a parametrisation for the m-dimensional manifold $M = \omega(G)$ embedded in $\mathbb{R}^n$ and of codimension $k = n - m$. The set $V(\delta)$ then consists of the projection to D of the complement in M of the set $\mathcal{D}(\delta, n) \cap M$ of points in M of Diophantine type (δ, n). The measure of the contributions to $|V(\delta)|$ arising from resonant sets $\{x \in \mathbb{R}^n \colon \langle \mathbf{q}\cdot x\rangle = 0\}$ which are transverse to M is

$$\ll \delta \sum_{\mathbf{q}\neq 0} |\mathbf{q}|^{-n-1} \ll \delta \sum_{r=1}^{\infty} \sum_{|\mathbf{q}|=r} r^{-n-1} \ll \delta \sum_{r=1}^{\infty} r^{-2} = O(\delta).$$

If an open subset of M coincides with part of the resonant set given by the hyperplane $R_{\mathbf{q}} = \{x \in \mathbb{R}^n \colon \mathbf{q}\cdot x = 0\}$ or if the curvature of M vanishes too often, then $|V(\delta)|$ will be large. If, however, M is 2-convex almost everywhere, these problems can be avoided and the estimates (7.17) and hence (7.16) hold on such manifolds. The proof has similarities with §2.5, and is done in some detail (but with a slightly different notation) in [88]. Kleinbock and Margulis' proof [139] that non-degenerate smooth manifolds are extremal raises the possibility that the estimate (7.16) might hold for such manifolds. As degenerate manifolds are locally flat, they can contain resonant planes, so that a better general result is out of the question.

7.6. Linearising diffeomorphisms

To say that a complex analytic diffeomorphism $f : \mathbb{C}^n \to \mathbb{C}^n$ can be linearised in a neighbourhood of a fixed point (which can be taken to be the origin) means that near the origin f is analytically (biholomorphically) conjugate to its linear part (Jacobian) $Df|_0 = A$ say. The linearising transformation ϕ is given by the solution to the functional equation

$$f \circ \phi = \phi \circ A$$

and is known as Schröder's equation when $n = 1$. Thus linearisation can regarded as obtaining a normal form and is analogous to diagonalising a matrix. Problems of small denominators, akin those discussed above, arise when the eigenvalues

$\alpha_1, \ldots, \alpha_n$ of A are close to being resonant in the sense that they are close to satisfying the equation

$$\alpha_k = \prod_{r=1}^{n} \alpha_r^{j_r}$$

for all $\mathbf{j} = (j_1, \ldots, j_n)$ with $|\mathbf{j}|_1 \geqslant 2$ and $j_r \geqslant 0$, $r = 1, \ldots, n$. Linearisation is well understood when $n = 1$ and the diffeomorphism $f : \mathbb{C} \to \mathbb{C}$ with $f(0) = 0$ can be linearised when $|(Df|_0)| = |f'(0)| \neq 1$. The interesting case when $|f'(0)| = 1$ is closely related (via lifts) to the conjugacy of a circle map to a rotation, discussed in §7.4 above, and necessary and sufficient conditions for the linearisation of a diffeomorphism $f \colon \mathbb{C} \to \mathbb{C}$ in terms of Bryuno numbers are known [118, 233, 234]. On the other hand, when $n \geqslant 2$, the problem of determining the precise class of functions which can be linearised is very difficult but Siegel [203, 204] established sufficient conditions on $Df|_0$ which guarantee the existence of a linearising transformation ϕ. The point $(\alpha_1, \ldots, \alpha_n)$ in $\mathbb{R}^n$ is said to be of *multiplicative type* (K, v) [8, p. 191] if

$$\left| \alpha_k - \prod_{r=1}^{n} \alpha_r^{j_r} \right| \geqslant K |\mathbf{j}|_1^{-v}$$

for all $\mathbf{j} \in (\mathbb{N} \cup \{0\})^n$ with $|\mathbf{j}|_1 \geqslant 2$. Siegel showed that if the vector $(\alpha_1, \ldots, \alpha_n)$ of eigenvalues of $Df|_0$ is of multiplicative type (K, v) for some positive numbers K and v, then f can be linearised in a neighbourhood (further details are in [8], [118], [167]). In order that Siegel's condition is not over restrictive, one takes $v > (n-1)/2$ as then for any $K > 0$ the set of points of multiplicative type (K, v) has full measure. However, the size of the neighbourhood of linearisation depends on v (it also depends on K but less significantly) and so it is desirable not to make v too large.

Let $\mathcal{E}_v$ denote the exceptional set of points in $\mathbb{C}^n$ (regarded as $\mathbb{R}^{2n}$) which for a given exponent v fail to be of multiplicative type (K, v) for any $K > 0$ and so fail to satisfy the conditions of Siegel's theorem. Then if $v > (n-1)/2$, $\mathcal{E}_v$ is null and

$$\dim \mathcal{E}_v = 2(n-1) + \frac{n+1}{v+1}. \tag{7.18}$$

This result is established by means of the map $(z_1, \ldots, z_n) \mapsto (e^{2\pi i z_1}, \ldots, e^{2\pi i z_n})$ which preserves the Hausdorff dimension and reduces the problem to sets with a simpler structure [94]. Multiplicative type is replaced by the more amenable notion of a point $z = x + iy$ in $\mathbb{C}^n$ being of *mixed additive type* (K, v), *i.e.*,

$$\max \left\{ |x_k - \sum_{r=1}^{n} j_r x_r|, \|y_k - \sum_{r=1}^{n} j_r y_r\| \right\} \geqslant K |\mathbf{j}|_1^{-v}$$

for each $\mathbf{j} \in (\mathbb{N} \cup \{0\})^n$ with $|\mathbf{j}|_1 \geqslant 2$ and each $k = 1, \ldots, n$. Given $v > 0$, the complementary set of points which are not of mixed additive type for any $K > 0$

is null. It is closely related to and has the same Hausdorff dimension as the set F_v say of points $(x, y) \in \mathbb{R}^{2n}$ such that

$$\max\{|\mathbf{j}\cdot x|, \|\mathbf{j}\cdot y\|\} < |\mathbf{j}|^{-v}$$

for infinitely many $\mathbf{j} \in \mathbb{Z}^n$. The upper bound for the Hausdorff dimension of F_v and hence of $\mathcal{E}_v$ follows from a straightforward and natural covering argument. The lower bound is obtained using ubiquity; for more details see [94] where a more general error function ψ is also considered.

Mixed additive type is also appropriate for the analysis of the related question of linearising periodic differential equations or, in other words, for finding the normal form of a vector field on $\mathbb{C}^n \times \mathbb{S}^1$ near a singular point. The periodic differential equation

$$\dot{z} = Bz + Q(z, t),$$

where B is an $n \times n$ complex matrix and $Q\colon \mathbb{C}^n \times \mathbb{S}^1 \to \mathbb{C}^n$ is analytic and has period 2π in t and satisfies

$$Q(0, t) = 0, \quad \frac{\partial Q_j(0, t)}{\partial z_k} = 0 \text{ for } 1 \leqslant j, k \leqslant n,$$

can be linearised to the form $\dot{\zeta} = B\zeta$ in a neighbourhood of 0 if the real and imaginary parts of the eigenvalues of B form a vector of mixed additive type (see [88, Theorem B], [204]). For each $v > 0$, the exceptional set of such vectors has the same Hausdorff dimension as $\mathcal{E}_v$ (7.18).

Similarly the autonomous differential equation

$$\dot{z} = Bz + Q(z), \tag{7.19}$$

where now $Q\colon \mathbb{C}^n \to \mathbb{C}^n$ is analytic and satisfies

$$Q(0) = 0, \quad \frac{\partial Q_j}{\partial z_k}(0) = 0 \text{ for } 1 \leqslant j, k \leqslant n,$$

can be linearised in a neighbourhoood of the origin if the eigenvalues $\beta_1, \ldots, \beta_n$ of B satisfy

$$|\beta_k - \mathbf{j}\cdot\beta| = |\beta_k - \sum_{r=1}^{n} j_r\beta_r| \geqslant K|\mathbf{j}|_1^{-v}$$

for each $\mathbf{j} \in (\mathbb{N} \cup \{0\})^n$ with $|\mathbf{j}|_1 \geqslant 2$ and each $k = 1, \ldots, n$, for some positive K, v (in which case $\beta \in \mathbb{C}^n$ is of *additive type* (K, v)). The set of points not of additive type (K, v) for given $v >$ and any $K > 0$ is null and has Hausdorff dimension $2(n-1) + n/(v+1)$ when $v > (n-2)/2$. Thus the exceptional set of points not of additive type (K, v) for any positive K, v is null and has Hausdorff dimension $2(n-1)$. Hence linearisation can be established if the eigenvalues are of multiplicative type; and so if the exceptional set of Hausdorff dimension $2(n-1)$ of eigenvalues not of multiplicative type is excluded.

The constant zero solution $z(t) = 0$ of (7.19) is called *(past and future) stable* if points near the origin stay near the origin under the evolution by (7.19) into the past and future. More precisely, every neighbourhood U of 0 contains a neighbourhood V of 0 such that $z(0) \in V$ implies that $z(t) \in U$ for all real t. By a theorem of Lyapunov, the eigenvalues of B must have non-positive real part for past stability and non-negative real part for future stability (strict inequality is a sufficient condition). Thus the zero solution is stable only if the eigenvalues are purely imaginary. But a remarkable theorem due to Carathéodory and Cartan [167, p. 22] tells us that the solution is stable if and only if B can be diagonalised and has purely imaginary eigenvalues $i\alpha_1, \dots, i\alpha_n$ and the vector field can be linearised analytically in a neighbourhood of 0. It thus follows from Siegel's theorem that the solution is stable if $(\alpha_1, \dots, \alpha_n) \in \mathbb{R}^n$ is a point of additive type (K, v) for some positive K, v. If $v > n-1$, than almost all vectors $(\alpha_1, \dots, \alpha_n)$ are of additive type (K, v) for some $K > 0$ and the Hausdorff dimension of the complementary exceptional set is $n - 1 + n/(v+1)$ [88]. Thus the zero solution is stable if the vector $(i\alpha_1, \dots, i\alpha_n)$ of eigenvalues of B lies outside a set of Hausdorff dimension $n-1$.

7.7. Diophantine approximation in hyperbolic space

The rationals $\mathbb{Q}$ can be interpreted as the orbit of the point at infinity under the linear fractional or Möbius transformation of the extended complex plane $\mathbb{C} \cup \{\infty\}$ in which

$$z \mapsto \frac{az+b}{cz+d}, \quad a, b, c, d \in \mathbb{Z}, \ ad - bc = 1,$$

i.e., as the orbit of ∞ under the modular group $\mathrm{SL}(2, \mathbb{Z})$ acting on points in the upper half plane. The set of limit points of $\mathbb{Q}$ is the real axis and $\mathrm{SL}(2, \mathbb{Z})$ is the discrete subgroup of $\mathrm{SL}(2, \mathbb{R})$, the group of orientation preserving Möbius transformations of the upper half plane to itself. These observations allow classical Diophantine approximation to be translated into a hyperbolic space setting by considering the orbit of a special point under the action of a Kleinian group G. When $(n+1)$-dimensional hyperbolic space is represented by the Poincaré ball model $(\mathbb{B}^{n+1}, \rho)$, where $\mathbb{B}^{n+1} = \{x \in \mathbb{R}^{n+1} \colon |x|_2 < 1\}$ is the unit ball and ρ is the hyperbolic metric derived from the differential $d\rho = |dx|_2/(1 - |x|_2^2)$, G is a discrete subgroup of the group of Möbius transformations which preserve orientation and $\mathbb{B}^{n+1}$. The elements of G are isometries of $(\mathbb{B}^{n+1}, \rho)$, *i.e.,* they preserve the hyperbolic metric ρ. Geodesics in this model are straight lines passing through the origin or arcs of circles intersecting the boundary $\mathbb{S}^n$ of the ball orthogonally. The alternative upper half space model $(\mathbb{H}^{n+1}, \rho)$ where $\mathbb{H}^{n+1} = \mathbb{R}^n \times \mathbb{R}^+$ and the metric ρ is derived from $d\rho = |dx|_2/x_{n+1}$ is equivalent to the ball model in that there is a Möbius transformation which maps $\mathbb{H}^{n+1}$ onto $\mathbb{B}^{n+1}$.

The relevant properties of Kleinian groups acting on hyperbolic space are now summarised; further details are given in [5], [20], [174]. The elements in G are compositions of two kinds of maps, namely similarity maps $x \mapsto Ax + b$, where

A is a conformal matrix (*i.e.,* a positive multiple of an orthogonal matrix) and $b \in \mathbb{R}^n$, and reflections in $\mathbb{S}^n$. As G is discrete, the necessarily countable orbit $G(x) = \{g(x) : g \in G\}$ of any point $x \in \mathbb{B}^{n+1}$ can accumulate only on the boundary $\mathbb{S}^n$. Denote by $\Lambda(G)$ the set of accumulation points of $G(x)$. Since the elements g of G are isometries with respect to the hyperbolic metric, the set $\Lambda(G)$ is the same for any $x \in \mathbb{B}^{n+1}$ and is called the *limit set* of G. The limit set is the smallest non-empty G-invariant subset of $\mathbb{S}^n$ and because G is a set of Möbius transformations, $\Lambda(G)$ consists of 0, 1, 2 or uncountably many points [20]. When $\Lambda(G)$ is finite, G is generated by a single Möbius transformation and is said to be *elementary.* Otherwise G is said to be *non-elementary* and in this case $\Lambda(G)$ is perfect. When $\Lambda(G) = \mathbb{S}^n$, G is said to be of the *first kind*; otherwise G is said to be of the *second kind.* Unless otherwise stated, the groups considered from now on will be non-elementary and *geometrically finite*, *i.e.,* there is a fundamental region in $\mathbb{B}^{n+1}$ which is a convex polyhedron with a finite number of faces. In the case $n = 1$, corresponding to Fuchsian groups acting on the unit disc $\mathbb{D}^2$ in $\mathbb{C}$, the notions of geometrically finite and finitely generated coincide. A geometrically finite group is of the second kind if and only if $\Lambda(G)$ is relatively null ($|\Lambda(G)|_{\mathbb{S}^n} = 0$); this follows at once from results of D. Sullivan [215] and P. Tukia [218] that for such G, $\dim \Lambda(G) < n$.

Each $g \in G$ is a Möbius transformation of $\mathbb{B}^{n+1}$. Being continuous, each g has a fixed point in the closed unit ball $\operatorname{cl} \mathbb{B}^{n+1}$ by the Brouwer fixed point theorem. Since g preserves $\operatorname{cl} \mathbb{B}^{n+1}$, g has at most two fixed points in $\operatorname{cl} \mathbb{B}^{n+1}$. A transformation g with just one fixed point on the boundary $\mathbb{S}^n$ is called *parabolic* and the point is called a parabolic fixed point. As an example, in $\mathbb{H}^{n+1}$ the point at infinity is a parabolic fixed point of the map $z \mapsto z+1$ in the modular group. A transformation g with two distinct fixed points on the boundary $\mathbb{S}^n$ is called *hyperbolic* and the point is called a hyperbolic fixed point. The fixed points in the interior of $\mathbb{B}^{n+1}$ are not of interest to us. A geometrically finite group G without parabolic elements is called *convex co-compact* since the action of G on the convex hull of $\Lambda(G)$ has a compact fundamental region in $\mathbb{B}^{n+1}$.

In the hyperbolic space setting the approximation of real numbers by rationals is replaced by approximating limit points of a Kleinian group by points in the orbit of a special or distinguished limit point $\mathfrak{p}$. More precisely, given $\xi \in \Lambda(G)$, we consider the quantity

$$|\xi - g(\mathfrak{p})|_2 \tag{7.20}$$

as g runs through G, where the special point $\mathfrak{p}$ is either a parabolic fixed point of G if G has such points or a hyperbolic fixed point otherwise. It can be seen that the orbit points $g(\mathfrak{p})$ as g runs through the group G play the role of the rationals and the limit set plays the role of the reals. We now need the notion of a 'denominator' in the hyperbolic space setting. Now the Jacobian $Dg|_x$ (often written $g'(x)$ in the literature) of each $g : \mathbb{B}^{n+1} \to \mathbb{B}^{n+1}$ at x is conformal, so that $(Dg|_x)/|\det(Dg|_x)|$ is orthogonal and $|\det(Dg|_x)|$ is the linear change of scale at x, depending only on

$|x|_2$. The quantity

$$\lambda_g = |\det(Dg|_0)|^{-1}$$

satisfies

$$\lambda_g = \frac{1}{2}\cosh \rho(0, g(0)) = \frac{1}{2} + \frac{|g(0)|_2^2}{1 - |g(0)|_2^2} \asymp e^{\rho(0,g(0))}$$

(see [174]), so that $\lambda_g \to \infty$ as $|g(0)| \to 1$, *i.e.,* as the orbit of the origin moves out towards the boundary of hyperbolic space. It turns out that not only is λ_g an appropriate analogue for the modulus of the denominator of a rational p/q in the classical setting but also properties of the Dirichlet series $\sum_{g\in G} \lambda_g^{-s}$ are connected with the Hausdorff dimension of $\Lambda(G)$. In fact the *exponent of convergence*

$$\delta(G) = \inf\{s > 0\colon \sum_{g\in G} \lambda_g^{-s} < \infty\}$$

of G is equal to $\dim \Lambda(G)$ [212, 215]. The appropriateness of λ_g as a 'denominator' of g can be seen clearly by considering the case $G = \mathrm{SL}(2,\mathbb{Z})$ and $\mathfrak{p} = \infty$ (the parabolic point at infinity). Each group element g satisfies $g(\infty) = p/q$ for some $p/q \in \mathbb{Q}$ and a straightforward calculation yields that λ_g is comparable to q^2 (in the upper half plane model $\lambda_g = |g'(i)|^{-1}$). The rich parallel theory of Diophantine approximation in hyperbolic space based on (7.20) and λ_g continues to develop and can only be sketched here. We begin with analogues of Dirichlet's theorem, where the precision of the approximation depends upon whether the distinguished point $\mathfrak{p}$ is parabolic or hyperbolic.

First, suppose that G has parabolic fixed points and let P be a complete set of such points inequivalent under G. Then there is a group constant $C > 0$ such that for each point $x \in \Lambda(G)$ and each integer $N \geqslant 2$, there exist a $\mathfrak{p} \in P$ and a $g \in G$ with $\lambda_g \leqslant N$ satisfying

$$|x - g(\mathfrak{p})|_2 \leqslant C/\sqrt{\lambda_g N}\,.$$

Otherwise let $\mathbf{y}, \mathbf{y}'$ be a conjugate pair of hyperbolic fixed points of G. Then there exists a $g \in G$ with $\lambda_g \leqslant N$ satisfying

$$|x - g(\mathfrak{p})|_2 \leqslant C/N\,,$$

where $\mathfrak{p}$ is one of $\mathbf{y}, \mathbf{y}'$. These statements were proved by S. J. Patterson in his seminal paper [177] for the case $n = 1$ (Fuchsian groups) and subsequently generalised to Kleinian groups [180, 212, 223]. Clearly, for each N the Dirichlet type results provide an economical covering of the limit set by balls centred at orbit points $g(\mathfrak{p})$ with $\mu(g) \leq N$. Moreover, by making use of the available geometric structure it is not difficult to see that the covering is of bounded multiplicity. This covering of the limit set plays a central role in establishing the hyperbolic space anologue of the classical metric Diophantine approximation theory.

The Dirichlet type theorems, together with various 'decoupling' results [159, 177, 223] which enable one to pick a specific distinguished point $\mathfrak{p}$, imply that there exists a group constant $C > 0$ for which given any $\xi \in \Lambda(G)$ there are infinitely many $g \in G$ satisfying

$$|\xi - g(\mathfrak{p})|_2 \leqslant C/\lambda_g \,.$$

When interpreted in the upper half plane model $\mathbb{H}^2$ and applied to the modular group $\mathrm{SL}(2,\mathbb{Z})$ with a careful analysis of the action, this gives Hurwitz's theorem [106]. The problem of determining to what extent the images of a distinguished point approximate an arbitrary limit point was raised by R. A. Rankin [188] and J. Lehner [148]. To some extent they were able to answer this and furthermore A. Beardon and B. Maskit [19, 20] have shown that the approximation properties of limit points reflect the geometry of the group.

The analogy with the classical theory can be taken further. Let $\psi\colon \mathbb{R}^+ \to \mathbb{R}^+$ converge monotonically to 0 at infinity. A point $\xi \in \Lambda(G)$ is called *ψ-approximable* if the inequality

$$|\xi - g(\mathfrak{p})|_2 < \psi(\lambda_g)/\lambda_g$$

holds for infinitely many $g \in G$. The set of ψ-approximable points will be denoted by $W(G,\mathfrak{p};\psi)$, *i.e.*,

$$W(G,\mathfrak{p};\psi) = \left\{\xi \in \Lambda(G) : |\xi - g(\mathfrak{p})|_2 < \frac{\psi(\lambda_g)}{\lambda_g} \text{ for infinitely many } g \in G\right\}$$

and is a counterpart of $\mathscr{S}(M;\psi)$ introduced in Chapter 1. The term $\psi(\lambda_g)/\lambda_g$ can be expressed in various equivalent ways and we follow Patterson [177] so that ψ corresponds to the approximation function ψ discussed earlier.

We now describe the hyperbolic space analogue of Khintchine's theorem. When G is of the first kind, $\Lambda(G) = \mathbb{S}^n$ and it makes sense to study the Lebesgue measure of $W(G,\mathfrak{p};\psi) \subseteq \Lambda(G)$, as one does for $\mathscr{S}(\mathbb{R}^n;\psi)$ in Khintchine's theorem. When G is of the second kind, however, $|\Lambda(G)|_{\mathbb{S}^n} = 0$ and so $W(G,\mathfrak{p};\psi)$ is a null set in $\mathbb{S}^n$ irrespective of ψ. The way forward is to work instead with *Patterson measure* μ [178], a non-atomic probability measure supported on $\Lambda(G)$ and the appropriate measure for the problem of metric Diophantine approximation in hyperbolic space (cf. §1.5.1). For groups of the first kind μ is normalised n-dimensional Lebesgue measure on $\mathbb{S}^n$ and for general groups without parabolic elements (*i.e.*, convex co-compact), μ is equivalent to $\delta(G)$-dimensional Hausdorff measure (recall that $\delta(G) = \dim \Lambda(G)$). When G is of the second kind with parabolic elements $\mathfrak{p}$, the Patterson measure has a fluctuating density which depends on the rank $r_{\mathfrak{p}}$ of $\mathfrak{p}$, where $r_{\mathfrak{p}}$ is the rank of a free abelian subgroup of the stabiliser $\{g \in G\colon g(\mathfrak{p}) = \mathfrak{p}\}$ of $\mathfrak{p}$ of finite index and $1 \leqslant r_{\mathfrak{p}} \leqslant n$. For groups of the first kind all parabolic fixed points are of maximal rank, *i.e.*, $r_{\mathfrak{p}} = \dim \Lambda(G) = n$; in general $r_{\mathfrak{p}} < 2\dim \Lambda(G)$ [18]. In the following counterpart of Khintchine's theorem, the Patterson measure of the set $W(G,\mathfrak{p};\psi)$ depends upon certain sums.

THEOREM 7.1. *Let G be a non-elementary geometrically finite group and let $\psi\colon [1/2,\infty) \to \mathbb{R}^+$ be decreasing and satisfy $\psi(2x) > c\,\psi(x)$ for some constant*

$c > 0$. *If G has a parabolic element $\mathfrak{p}$ of rank $r_\mathfrak{p}$, then $\mu(W(G,\mathfrak{p};\psi)) = 0$ or 1 according as the sum*

$$\sum_{k=1}^{\infty} \psi(K^k)^{\,2\delta(G)-r_\mathfrak{p}}$$

converges or diverges for some $K > 1$.

If G has no parabolic elements, then $\mu(W(G,\mathfrak{p};\psi)) = 0$ or 1 for any hyperbolic element $\mathfrak{p}$ according as the sum

$$\sum_{k=1}^{\infty} \psi(K^k)^{\delta(G)}$$

converges or diverges for some $K > 1$.

It was established for groups of the first kind (where Patterson measure is comparable to Lebesgue measure) by Patterson [177, 180]. For groups of the first kind with parabolic elements, Sullivan [215] gave a geometric proof and a dynamic interpretation of the theorem which enabled him to prove a 'logarithmic law for geodesics', to be discussed later. Theorem 7.1 was proved for groups without parabolic elements in [86] and for groups with parabolic elements in [212].

The differences between Theorem 7.1 and the classical result arise from the hyperbolic metric; the behaviour of the sums is essentially the same. Apart from the mild growth condition on ψ, Theorem 7.1 reduces to Khintchine's theorem when interpreted on the upper half plane model and applied to the modular group. Convergence follows readily enough from Cantelli's lemma together with elementary counting arguments. To establish the more difficult divergence case, a G-invariant subset of $W(G,\mathfrak{p};\psi)$ satisfying a pairwise quasi-independence property is constructed. This property together with the divergence of the sums implies that the invariant subset has positive Patterson measure and the result follows from the action of G being ergodic on $\Lambda(G)$ with respect to the measure. This line of argument is in the same spirit as the classical case.

For groups with parabolic elements, Theorem 7.1 can be used to obtain a general version of Sullivan's logarithmic law and to give another proof of the important result that the exponent of convergence $\delta(G)$ is equal to $\dim \Lambda(G)$ [212]. In fact the theorem can be also used to establish the deeper result result the Hausdorff measure of $\Lambda(G)$ is infinite with respect to particular dimension functions which are roughly speaking close to $r^{\delta(G)}$ [212, 219].

Having introduced the notion of ψ-approximable limit points, we briefly discuss the hyperbolic space analogue of badly approximable numbers. A point β in $\Lambda(G)$ is said to be *badly approximable with respect to* $\mathfrak{p}$ if there exists a positive constant $K = K(\beta)$ such that for all $g \in G$

$$|\beta - g(\mathfrak{p})|_2 \geqslant K/\lambda_g\,.$$

It follows from the hyperbolic space counterpart of Khintchine's theorem discussed above that the set $\mathfrak{B}(G,\mathfrak{p})$ of badly approximable limit points is of Patterson measure zero. However, the set is of full Hausdorff dimension $\dim \Lambda(G)$, so that the analogue of Jarník's theorem (that $\dim \mathfrak{B} = 1$ [125]) holds [50, 104, 177, 180]. In fact, J.-L. Fernandez and M. V. Melián [104] have shown that for any non-elementary Fuchsian group G (including infinitely generated ones) the dimension of $\mathfrak{B}(G, \mathfrak{p})$ is equal to the exponent of convergence $\delta(G)$ and in higher dimensions, C. J. Bishop and P. W. Jones [50] have shown that the same result holds for finitely generated Kleinian groups. When interpreted on $\mathbb{H}^2$ and applied to the modular group, all these results reduce to that of Jarník. Badly approximable points will not be discussed in detail; further information and references can be found in the works cited above.

Turning to Hausdorff dimension results for the set $W(G, \mathfrak{p}; \psi)$, we begin by considering the hyperbolic space analogue of the Jarník-Besicovitch theorem. For the rest of this section, let $v \geqslant 0$. When $\psi(x) = x^{-v}$, write $W(G, \mathfrak{p}; \psi) = W_v(G, \mathfrak{p})$; by analogy with the classical case, its elements are called *very well approximable* when $v > 0$. Then $W_v(\mathrm{SL}(2, \mathbb{Z}), \infty)$ essentially corresponds to the set $\mathscr{K}_{2v+1}(\mathbb{R})$; for any $\varepsilon > 0$,

$$W_v(\mathrm{SL}(2, \mathbb{Z}), \infty) \subset \mathscr{K}_{2v+1}(\mathbb{R}) \subset W_{v-\varepsilon}(\mathrm{SL}(2, \mathbb{Z}), \infty).$$

Analogues of the Jarník-Besicovitch theorem were proved independently for groups of the first kind by Melián and Pestana [158] using 'well-distributed' systems (see the Notes on §5.2) and by Velani [222, 223] who used ubiquity (see §5.3). The result was proved for convex co-compact groups in [86] by extending the notion of ubiquity from Lebesgue measure to Patterson measure. The complete analogue for the parabolic case of the Jarník-Besicovitch theorem was established by Hill and Velani [123] who constructed a Cantor type subset of $W_v(G, \mathfrak{p})$ on which the fluctuations of Patterson measure were sufficiently well controlled. This in turn enabled them to make optimal use of the mass distribution principle (§3.5.7).

THEOREM 7.2. *Let G be a non-elementary geometrically finite group. Suppose G has a parabolic element $\mathfrak{p}$. Then for $v \geqslant 0$*

$$\dim W_v(G, \mathfrak{p}) = \min\left\{\frac{\delta(G) + r_{\mathfrak{p}} v}{2v+1}, \frac{\delta(G)}{v+1}\right\}.$$

Otherwise G is convex co-compact and $\dim W_v(G, \mathfrak{p}) = \delta(G)/(v+1)$ *for $v \geqslant 0$.*

Thus if $r_{\mathfrak{p}} \geqslant \delta(G)$ or if $v + 1 \geqslant \delta(G)/r_{\mathfrak{p}}$, then $\dim W_v(G, \mathfrak{p}) = \delta(G)/(v+1)$. When G is a geometrically finite group of the first kind with parabolic points, $\dim W_v(G, \mathfrak{p}) = n/(v+1)$ when $v \geqslant 0$; the Jarník-Besicovitch theorem can be recovered on putting $G = \mathrm{SL}(2, \mathbb{Z})$ and $\mathfrak{p} = \infty$. When G is of the first kind or is convex co-compact, the theorem can be extended to the more general set $W(G, \mathfrak{p}; \psi)$, with dimension a function of the lower order at infinity of $1/\psi$, as in the classical case [224].

When $\mathfrak{p}$ is a parabolic fixed point of G, the sets $\mathfrak{B}(G, \mathfrak{p})$ and $W(G, \mathfrak{p}; \psi)$ have a beautiful dynamical interpretation in terms of the rate at which geodesics make excursions into cuspidal ends of an associated manifold M. In order to explain this, some notation is needed. When G is geometrically finite, the quotient or identification space $M = \mathbb{B}^{n+1}/G$ obtained by identifying equivalent points in $\mathbb{B}^{n+1}$ under the action of G is an n-dimensional manifold. When G is a Fuchsian group, the manifold is a Riemann surface of constant negative curvature, in particular the modular surface $\mathbb{H}^2/\mathrm{SL}(2,\mathbb{Z})$ is the punctured torus. The manifold M can be decomposed into a disjoint union of a compact part with a finite number of exponentially 'narrowing' cuspidal ends (corresponding to a set of inequivalent parabolic fixed points of G) and 'exploding' ends or funnels (corresponding to the free faces of a convex fundamental polyhedron for G).

For each $x \in M$, let $\mathbb{S}(x)$ be the unit sphere in the tangent space T_xM at x, and given a vector u in $\mathbb{S}(x)$, let γ_u be the geodesic emanating from x in the direction u. For each $t \in \mathbb{R}^+$, let $\gamma_u(t)$ denote the point reached after travelling time t along γ_u and let $\mathrm{dist}(x, \gamma_u(t))$ be the distance from x to $\gamma_u(t)$ induced on M by the hyperbolic metric ρ in $\mathbb{B}^{n+1}$. By considering the segment joining a point in $\mathbb{B}^{n+1}$ to the point $\xi \in \Lambda(G)$ and its projection to M, it can be seen that the approximation properties of ξ are related to the behaviour of the corresponding geodesic from x in the manifold M. The limit points of G correspond to those directions u in $\mathbb{S}(x)$ for which the geodesics γ_u do not end up in the funnels of M. The set $\mathfrak{B}(G,\mathfrak{p})$ corresponds to bounded geodesics on the manifold M. More precisely, it corresponds exactly to the set of u in $\mathbb{S}(x)$ for which $\gamma_u(t)$ remains bounded for all time t (see for example [50], [104]). Thus the dimension results for $\mathfrak{B}(G,\mathfrak{p})$ imply that the set of bounded geodesics on M is of dimension $\delta(G)$. On the other hand, the set $W(G,\mathfrak{p};\psi)$ can be interpreted in terms of the rate of excursions by geodesics into the cuspidal ends of the manifold (the rate being governed by the function ψ); *i.e.,* in terms of geodesics which persistently enter a 'shrinking' neighbourhood of the cusp corresponding to $\mathfrak{p}$. In particular, the Khintchine type theorem for $W(G,\mathfrak{p};w)$ (Theorem 7.1 above) leads to a general 'logarithmic law for geodesics'. For μ-almost all $u \in \mathbb{S}(x)$

$$\limsup_{t\to\infty} \frac{\mathrm{dist}(x,\gamma_u(t))}{\log t} = \frac{1}{2\delta(G) - r_{\max}},$$

where $r_{\max}$ denotes the maximum rank of the parabolic fixed points of G [212]. For groups of the first kind ($\delta(G) = n = r_{\max}$), one recovers Sullivan's logarithmic law [215]. S. G. Dani has made profound generalisations of badly and very well approximable numbers in terms of 'bounded' and 'divergent' trajectories on quotient and homogeneous spaces (see for example [65], [66], [67], [68], [69]).

For the sake of simplicity assume that G is of the first kind. Then the dimension results for $W(G,\mathfrak{p};\psi)$ can be interpreted in the following way. Let $f\colon \mathbb{R}^+ \to \mathbb{R}^+$ be such that $t - f(t)$ is strictly increasing. Then

$$\dim\left\{u \in \mathbb{S}(x)\colon \limsup_{t\to\infty} \frac{\mathrm{dist}(x,\gamma_u(t))}{f(t)} \geq 1\right\} = n\left(1 - \limsup_{t\to\infty}\frac{f(t)}{t}\right).$$

In particular if $f(t) = (1-\beta)t$, $0 \leqslant \beta < 1$, then the dimension is $n\beta$ and is a dynamical interpretation of the result $\dim W_v(G;\mathfrak{p}) = n/(v+1)$, $v+1 = 1/\beta > 1$

([224], the notation has been changed slightly). For groups of the second kind, the corresponding result can be found in [123]. Thus the Jarník-Besicovitch theorem gives information about how many geodesics enter a shrinking neighbourhood of the cusp at infinity of the modular surface.

This leads to the more general idea of a *shrinking target* introduced by Hill and Velani in [121] and further exploited in [122] and [120]. Let X be a metric space which carries a Borel probability measure μ. Let $T\colon X \to X$ and $\varrho\colon \mathbb{R}^+ \to \mathbb{R}^+$ be functions and suppose T preserves measure and is ergodic. Then the set

$$W(a,\varrho) = \{x \in X\colon T^q(x) \in B(a,\varrho(q)) \text{ for infinitely many } q \in \mathbb{N}\},$$

where $B(a,\varepsilon)$ is a ball of radius ε in X, is an analogue of the sets $W(G,\mathfrak{p};\psi)$ and $\mathscr{S}(\mathbb{R}^n;\psi)$. A natural question is how the size of $W(a,\varrho)$ depends on ϱ. By Birkhoff's ergodic theorem [49, 181], if $\varrho(q) = \varrho_0 > 0$ for q sufficiently large and $B(a,\varrho)$ has positive μ-measure, the set

$$\{x \in X\colon T^q(x) \in B(a,\varrho_0) \text{ for infinitely many } q \in \mathbb{N}\}$$

has full μ-measure; *i.e.*, the forward orbit of almost all points falls into the ball infinitely often. The above analysis implies that the complement of $W(a,\varrho_0)$ in X, which consists of points whose forward orbits land in the ball only a finite number of times, is of zero μ-measure. This set corresponds to the set of badly approximable points in a dynamical system which has been studied extensively by various authors (details can be found in [2], [4], [3], [69], [220]).

On the other hand, when $\varrho(q) \to 0$ as $q \to \infty$, points in $W(a,\varrho)$ have trajectories which hit a shrinking ball or target infinitely often and so are called *ϱ-approximable.* The points in the backward orbit of the point a in X are resonant points corresponding to the rationals in $\mathscr{S}_v(\mathbb{R})$ and orbit points in $W(G,\mathfrak{p};\psi)$. This framework enables ideas developed for Diophantine approximation in the hyperbolic space setting together with ideas from ergodic theory to be used to analyse the structure of a variety of apparently quite unrelated sets in complex dynamics and other dynamical systems. For instance, take T to be a rational map of degree ≥ 2 on the Riemann sphere and take $X = J(T)$, the Julia set of T; from now on we write $\delta(T) = \dim J(T)$. Choosing $\varrho(q) = |(T^q)'(x)|^{-v}$ (so that $\varrho(q)$ depends on x) gives the 'local' set

$$W_v^\bullet(a) = \{x \in J(T)\colon T^q(x) \in B(a, |(T^q)'(z)|^{-v}) \text{ for infinitely many } q \in \mathbb{N}\}$$

and choosing $\varrho(q) = \exp(-vq)$ gives the 'global' set

$$W_v(a) = \{x \in J(T)\colon T^q(x) \in B(a, e^{-vq}) \text{ for infinitely many } q \in \mathbb{N}\}.$$

The local set $W_v^\bullet(a)$ is the analogue of $W_v(G,\mathfrak{p})$ when G is a Kleinian group and when T is the familiar continued fraction map, given by $T(x) = \{1/x\}$ for $x \in (0,1]$ and $T(0) = 0$, the set $W_v^\bullet(0)$ is directly related to $\mathscr{S}_v(\mathbb{R})$ [122].

When T is an expanding rational map, $\dim W_v^\bullet(a) = \delta(T)/(v+1)$ [121]. This is evidently an analogue of Theorem 7.2 when the distinguished point $\mathfrak{p}$ is hyperbolic and can be established using a form of ubiquity adapted to Sullivan's

$\delta(T)$-conformal measure [213]. When $J(T)$ contains rationally indifferent periodic points a but no critical points (in which case T is usually referred to as a parabolic rational map), the approximation properties are very different and have similarities with the case of Kleinian groups with parabolic elements. Let a be a rationally indifferent periodic point of $J(T)$ and ℓ denote the number of petals asociated with the flower around a (the terminology is explained in [21]). Then for $v \geqslant 0$,

$$\dim W_v^\bullet(a) = \min\left\{\frac{\delta(T)+v\ell}{v(\ell+1)+1}, \frac{\delta(T)}{v+1}\right\},$$

which is analogous to Theorem 7.2 when $\mathfrak{p}$ is parabolic [119]. The technical problems caused by the indifferent periodic points in $J(T)$ are overcome by the use of further ideas from ergodic theory.

In the case of expanding maps, the approach used in [121] for the local set $W_v^\bullet(a)$ only yields a partial result on the dimension of the global set $W_v(a)$. To calculate $\dim W_v(a)$, consider the Hölder continuous function $f\colon J(T) \to \mathbb{R}^+$ satisfying $f(x) > \log|T'(x)|$. The set $D(a,f)$ of points in $J(T)$ which lie in the ball

$$B\left(y, \exp(-\sum_{j=0}^{q-1} f(T^j(y)))\right)$$

for infinitely many pairs (y,q) with $T^q(y) = a$ corresponds in essence to the set $W(a,\varrho)$ expressed in terms of the pre-images of a. It has Hausdorff dimension s_0, where $s_0 = s_0(f,T)$ is the unique solution of the equation $P(T,-sf) = 0$ and P is the topological pressure on the Julia set [122]. This is an extension of the Bowen-Manning-McCluskey formula $P(T, -\delta(T)\log|T'|) = 0$. The proof involves forcing an f-conformal measure [72] (a generalisation of the δ-conformal measures of Patterson and Sullivan) onto a Cantor type subset and using the mass distribution principle. By choosing f appropriately, the Hausdorff dimensions of $W_v^\bullet(a)$ and $W_v(a)$ can be obtained, giving analogues of the Jarník-Besicovitch theorem. In particular, the set $W_v(a)$ corresponds to letting $f(x) = \log|T'(x)| + v$, giving $\dim W_v(a) = s_0$ where $P(T, -s_0 \log|T'|) = s_0\, v$.

These techniques have been applied to expanding Markov maps T of the unit interval (they include the continued fraction map). The theory has been extended to higher dimensional tori and to maps which are multiplication by integer matrices modulo $\mathbb{Z}^n$ [120]. The very well approximable sets associated with a given dynamical system have rather unexpected links with exceptional sets arising from points in the phase space which have 'badly behaved' ergodic averages and with multifractal spectra [102, 122].

7.8. Notes

§**7.1** Diophantine approximation appears to be have been used by the ancient Greeks to design sophisticated gearing [185]. If David Fowler's radical but persuasive proposal [107] that the ancient Greeks defined ratio using the Euclidean algorithm is correct, then it is likely that Diophantine approximation techniques were used. Continued fractions were used by Huygens and others in the 17th and 18th centuries to construct gear ratios in clocks and orreries.

The first successful attack on a problem of small denominators (apart possibly from Dirichlet who remarked to Kronecker that he had found a way of approximating successively the solutions to the N-body problem [167, p. 8]) was due to C. L. Siegel, who treated the question of the stability of the differential equation $\dot{z} = f(z)$, where $f\colon \mathbb{C}^n \to \mathbb{C}^n$ is analytic near the origin (discussed above in §7.6) in [203].

There is a difference in point of view: 'very well approximable' numbers of Diophantine approximation are the source of the problems of convergence and so in this sense are 'bad', while the 'badly approximable' numbers are good for convergence – the golden ratio is the best and quadratic irrationals are sometimes called 'noble numbers'.

§7.2 The null sets $E_\sigma = \bigcap_{K\in\mathbb{Q}^+}\{\xi \in \mathbb{R}\colon |\xi - p/q| < Kq^{-2-\sigma}$ for some $p/q \in \mathbb{Q}\}$ and E are of second Baire category while their complements (and so $\mathcal{B}$) are of full measure but of first Baire category.

§7.4 Orientation preserving, pairwise commuting circle diffeomorphisms are simultaneously conjugate to rotations if each is close to a rotation and their rotation numbers cannot be approximated simultaneously too closely by rationals with the same denominator [169], for higher dimensions see [109].

A perfect, nowhere dense set is compact, totally disconnected and without isolated points; it is also homeomorphic to the Cantor set.

§7.3 Very small exceptional sets occur in obtaining normal forms for pseudo-differential operators on the torus [78]. The analysis involves deeper Hausdorff measure results from [79].

Various kinds of problems of small denominators, including those of the type $\|q\xi_1\|^{n_1}\dots\|q\xi_m\|^{n_m}$ treated in [81], arise in badly posed boundary problems [46, 186].

§7.5.2 J. Moser [166, 168] reduced the requirement of analyticity to several hundred derivatives and Rüssmann [190] reduced the differentiability substantially using majorisation arguments (see also [233]).

It is known that for $n = 2$ and under certain conditions, $|I(t) - I(0)|_2 < \alpha(\varepsilon)$ where $\alpha(\varepsilon) \to 0$ as $\varepsilon \to 0$ for all $t \in \mathbb{R}$. For $n \geqslant 3$, N. N. Nehoroŝev has shown that for real analytic, near integrable Hamiltonian systems $H_0(I) + \varepsilon H(I, \Theta)$, when H_0 obeys certain 'steepness' properties, the time for which $|I(t) - I(0)|_2 < \alpha(\varepsilon)$ grows exponentially with respect to $1/\varepsilon$ [171, 172]. P. Lochak has obtained similar results when H_0 is quasi-convex (the level surface of the energy $H_0(I) = E$ is strictly convex for an interval of energies E) and $\alpha(e) = c\varepsilon^b$ [150].

A set involving systems of linear forms and essentially a more general form of (7.14) is in [73].

§7.7 All Riemann surfaces arise by forming the quotient space with respect to a discontinous group action [20]. For an account of the connection between geodesics on the modular surface $\mathbb{H}^2/\mathrm{SL}(2,\mathbb{Z})$ and Farey sequences, see [202].

A measure ν on J is said to be f-conformal (with respect to T) for a Hölder continuous function $f\colon J(T) \to \mathbb{R}^+$ if $\nu(TA) = \int_A e^{f(x)}\,d\nu(x)$ for every Borel set $A \subset J$ such that $T|_A$ is injective. When $f = \delta \log|T'|$, the measure is said to be δ-conformal.

References

[1] A. G. Abercrombie, The Hausdorff dimension of some exceptional sets of p-adic matrices, *J. Number Th.* **53** (1995), 311–341.

[2] A. G. Abercrombie and R. Nair, The ergodic theory of shrinking targets, *Ergod. Th. Dyn. Sys.*, to appear.

[3] ———, An exceptional set in the ergodic theory of Markov maps on the interval, *Proc. Lond. Math. Soc.* **75** (1997), 221–240.

[4] ———, An exceptional set in the ergodic theory of rational maps on the Riemann sphere, *Ergod. Th. Dyn. Sys.* **17** (1997), 253–267.

[5] L. V. Ahlfors, *Möbius transformations in several dimensions*, Lecture Notes, School of Mathematics, University of Minnesota, 1954.

[6] V. I. Arnol'd, Small denominators and problems of stability of motion in classical and celestial mechanics, *Usp. Mat. Nauk* **18** (1963), 91–192, English transl. in *Russian Math. Surveys* **18** (1963), 85–191.

[7] ———, *Mathematical methods of classical mechanics*, Springer-Verlag, 1978, Translated by K. Vogtmann and A. Weinstein.

[8] ———, *Geometrical methods in ordinary differential equations*, Springer-Verlag, 1983, Translated by J. Szücs.

[9] V. I. Arnol'd, V. V. Kozlov, and A. I. Neishtadt, *Mathematical aspects of classical and celestial mechanics*, Encyclopaedia of Mathematical Sciences, vol. 3, Dynamical Systems III, Springer-Verlag, 1980, Translated by A. Jacob.

[10] A. Baker, On a theorem of Sprindžuk, *Proc. Roy. Soc. Series A* **292** (1966), 92–104.

[11] ———, *Transcendental number theory*, Cambridge University Press, 1974.

[12] ———, *A concise introduction to number theory*, Cambridge University Press, 1984.

[13] A. Baker and W. M. Schmidt, Diophantine approximation and Hausdorff dimension, *Proc. Lond. Math. Soc.* **21** (1970), 1–11.

[14] R. C. Baker, Metric Diophantine approximation on manifolds, *J. Lond. Math. Soc.* **14** (1976), 43–48.

[15] ———, Sprindžuk's theorem and Hausdorff dimension, *Mathematika* **23** (1976), 184–197.

[16] ———, Dirichlet's theorem on Diophantine approximation, *Math. Proc. Cam. Phil. Soc.* **83** (1978), 37–59.

[17] V. I. Bakhtin, Diophantine approximation on images of maps, *Dokl. Akad. Nauk BSSR* **35** (1991), 398–400.

[18] A. F. Beardon, The exponent of convergence of Poincaré series, *Proc. Lond. Math. Soc.* **18** (1968), 461–483.

[19] ———, Limit points of Kleinian groups and finite sided fundamental polyhedra, *Acta Math.* **132** (1974), 1–12.

[20] ———, *The geometry of discrete groups*, Springer-Verlag, 1983.

[21] ———, *Iteration of rational functions*, Springer-Verlag, 1991.

[22] V. V. Beresnevich, On approximation of real numbers by real algebraic numbers, *Acta Arith.*, to appear.

[23] ———, Exact metric theorems in the theory of Diophantine approximation, preprint 10, pp. 1–20, Inst. Math., Belarus Acad. Sc., 1997.

[24] ———, A Khintchine-Groshev type theorem for convergence on manifolds, preprint, University of York, 1999.

[25] V. V. Beresnevich and V. I. Bernik, On a metrical theorem of W. Schmidt, *Acta Arith.* **75** (1996), 219–233.

[26] V. V. Beresnevich, V. I. Bernik, H. Dickinson, and M. M. Dodson, The Khintchine-Groshev theorem for planar curves, *Proc. Roy. Soc. Lond. A*, to appear.

[27] V. V. Beresnevich, V. I. Bernik, and M. M. Dodson, Inhomogeneous non-linear Diophantine approximation, *Papers in honour of V. G. Sprindžuk's 60th birthday*, Inst. Math., Belarus Acad. Sc., 1997, pp. 13–20.

[28] V. I. Bernik, Asymptotic number of solutions of some systems of Diophantine inequalities, *Mat. Zametki* **11** (1972), no. 6, 619–623.

[29] ———, Asymptotic number of solutions of certain systems of inequalities in the theory of Diophantine approximation of dependent quantities, *Vestsī Akad. Navuk BSSR, Ser. Fīz.-Mat.* **1** (1973), 10–17.

[30] ———, Induced extremal surfaces, *Mat. Sbornik* **103** (1977), 480–489.

[31] ———, On the exact order of approximation of almost all points on the parabola, *Mat. Zametki* **26** (1979), 657–665.

[32] ———, A metric theorem on the simultaneous approximation of zero by the values of integral polynomials, *Izv. Akad. Nauk SSSR, Ser. Mat.* **44** (1980), no. 1, 24–45, English transl. in *Math. USSR Izvest.*, **16** (1983), 21–40.

[33] ———, An application of Hausdorff dimension in the theory of Diophantine approximation, *Acta Arith.* **42** (1983), 219–253, English transl. in *Amer. Math. Soc. Transl.* **140** (1988), 15–44.

[34] ———, A proof of Baker's conjecture in the theory of transcendental numbers, *Dokl. Akad. Nauk SSSR, Ser. Mat.* **277** (1984), no. 5, 1036–1039.

[35] ———, Applications of measure theory and Hausdorff dimension to the theory of Diophantine approximation, *New advances in transcendence theory* (A. Baker, ed.), Cambridge University Press, 1988, pp. 25–36.

[36] ———, On the exact order of approximation of zero by values of integral polynomials, *Acta Arith.* **53** (1989), 17–28 (Russian).

[37] V. I. Bernik, H. Dickinson, and M. M. Dodson, A Khintchine-type version of Schmidt's theorem for planar curves, *Proc. Roy. Soc. Lond. A* **454** (1998), 179–185.

[38] V. I. Bernik, H. Dickinson, and J. Yuan, *Inhomogeneous Diophantine approximation on polynomial curves in* $\mathbb{Q}_p$, preprint, University of York, 1997.

[39] V. I. Bernik and I. R. Dombrovsky, Effective estimates of measure of sets defined by Diophantine conditions, Number theory and analysis (Z. D. Kudryavtsov, ed.), vol. 207, *Trudy Mat. Inst. Stekl.*, Nauka, 1994, Proceedings of International Conference for the centenary of the birth of I. M. Vinogradov, Moscow 1991, pp. 35–41.

[40] V. I. Bernik and Kovaleskaya E. I., Diophantine approximation on n-dimensional manifolds in $\mathbb{R}^{2n}$, *Dokl. Bel.* **12** (1990), 1061–1064.

[41] V. I. Bernik, D. Y. Kleinbock, and G. A. Margulis, *Khintchine-type theorems for Diophantine approximation on manifolds*, preprint, 1999.

[42] V. I. Bernik and Y. I. Melnichuk, On integer polynomials of a p-adic variable with small norm in the disc, *Proc. Lvov Polytech. Inst.* **182** (1985), 63–64.

[43] ———, *Diophantine approximation and Hausdorff dimension*, Akad. Nauk BSSR, 1988.

[44] V. I. Bernik and I. L. Morotskaya, Diophantine approximation in $\mathbb{Q}_p$ and Hausdorff dimension, *Vestsī Akad. Navuk BSSR Ser. Fīz.-Mat.* (1986), no. 3, 3–9, 123.

[45] V. I. Bernik and N. A. Pereverseva, The method of trigonometric sums and lower estimates of Hausdorff dimension, *New Trends in Probability and Statistics, Analytic and Probabilistic Methods in Number Theory: Proceedings of the International Conference in Honour

of J. Kubilius, Palanga, Lithuania, September 1991 (F. Schweiger and E. Manstavičius, eds.), vol. 2, 1992, pp. 75–81.

[46] V. I. Bernik and B. I. Ptashnik, A boundary value problem for a system of partial differential equations with constant coefficients, *Diff. Uravneniya* **16** (1980), 273–291, English transl. in Diff. Eqns. **16** (1980), 176–180.

[47] A. S. Besicovitch, On linear sets of points of fractional dimensions, *Math. Ann.* **101** (1929), 161–193.

[48] ______, Sets of fractional dimensions (IV): on rational approximation to real numbers, *J. Lond. Math. Soc.* **9** (1934), 126–131.

[49] P. Billingsley, *Ergodic theory and information*, John Wiley, 1965.

[50] C. J. Bishop and P. W. Jones, Hausdorff dimension and Kleinian groups, *Acta Math.* **111** (1997), 1–39.

[51] E. Borel, Sur un problème de probabilités aux fractions continues, *Math. Ann.* **72** (1912), 578–584.

[52] I. Borosh and A. S. Fraenkel, A generalisation of Jarník's theorem, *Indag. Math.* **34** (1972), 193–201.

[53] J. D. Bovey and M. M. Dodson, The fractional dimension of sets whose simultaneous rational approximations have errors with a small product, *Bull. Lond. Math. Soc.* **10** (1978), 213–218.

[54] ______, The Hausdorff dimension of systems of linear forms, *Acta Arith.* **45** (1986), 337–358.

[55] A. D. Bryuno, On conditions for non-degeneracy in Kolmogorov's theorem, *Dokl. Akad. Nauk SSSR* **322** (1992), 1028–1032, English transl. in *Soviet Math. Dokl.* **45** (1992), 221–225.

[56] C. Carathéodory, Über das lineare Mass von Punktmengen, eine Verallgemeinerung des Längenbegriffs, *Gött. Nachr.* (1914), 404–226.

[57] L. Carleson, *Selected problems in exceptional sets*, van Nostrand, 1967.

[58] J. W. S. Cassels, Some metrical theorems in Diophantine approximation. V: on a conjecture of Mahler, *Proc. Cam. Phil. Soc.* **47** (1951), 18–21.

[59] ______, *An introduction to Diophantine approximation*, Cambridge Tracts in Math. and Math. Phys., vol. 99, Cambridge University Press, 1957.

[60] ______, *An introduction to the geometry of numbers*, Die Grundlehren der Mathematischen Wissenschaften, vol. 99, Springer, 1959.

[61] ______, *Local fields*, Cambridge University Press, 1986.

[62] I. Chavel, *Riemannian geometry: a modern introduction*, Cambridge Tracts in Math., vol. 108, Cambridge University Press, 1995.

[63] K. L. Chung, *A course in probability theory*, 2nd ed., Academic Press, 1974.

[64] E. F. Collingwood, Emile Borel, *J. Lond. Math. Soc.* **34** (1959), 488–512.

[65] S. G. Dani, On orbits of unipotent flows on homogeneous spaces, *Ergod. Th. Dyn. Sys.* **4** (1984), 25–34.

[66] ______, Divergent trajectories of flows on homogeneous spaces and homogeneous Diophantine approximation, *J. reine angew. Math.* **359** (1985), 55–89.

[67] ______, Bounded orbits of flows on homogeneous spaces, *Comm. Math. Helv.* **61** (1986), 636–660.

[68] ______, Orbits of horospherical flows, *Duke Math. J.* **53** (1986), 177–188.

[69] ______, On badly approximable numbers, Schmidt games and bounded orbits of flows, *Number theory and dynamical systems* (M. M. Dodson and J. A. G. Vickers, eds.), LMS Lecture Note Series, vol. 134, Cambridge University Press, 1987, pp. 69–86.

[70] H. Davenport and W. M. Schmidt, Dirichlet's theorem on Diophantine approximation, *Inst. Alt. Mat. Symp. Math.* **4** (1970), 113–132.

[71] ———, Dirichlet's theorem on Diophantine approximation II, *Acta Arith.* **16** (1970), 413–424.

[72] M. Denker and M. Urbánski, Geometric measures for parabolic rational maps, *Ergod. Th. Dyn. Sys.* **12** (1992), 53–66.

[73] H. Dickinson, The Hausdorff dimension of systems of simultaneously small linear forms, *Mathematika* **40** (1993), 367–374.

[74] ———, The Hausdorff dimension of sets arising in Diophantine approximation, *Acta Arith* **53** (1994), 133–140.

[75] ———, A remark on a theorem of Jarník, *Glasgow Math. J.* **39** (1997), 233–236.

[76] H. Dickinson and M. M. Dodson, *Extremal manifolds and Hausdorff dimension*, Duke Math. J., to appear.

[77] H. Dickinson, M. M. Dodson, and Jin Yuan, Hausdorff dimension and p-adic Diophantine approximation, *Indag. Math., N.S.* **9** (1998), 1–12.

[78] H. Dickinson, T. Gramchev, and M. Yoshino, First order pseudodifferential operators on the torus: normal forms, Diophantine approximation and global hypoellipticity, *Ann. Univ. Ferrara, Sez. VII - Sc. Mat.* **61** (1995), 51–64.

[79] H. Dickinson and S. L. Velani, Hausdorff measure and linear forms, *J. reine angew. Math.* **490** (1997), 1–36.

[80] M. M. Dodson, A note on the Hausdorff-Besicovitch dimension of systems of linear forms, *Acta Arith.* **44** (1985), 87–98.

[81] ———, Star bodies and Diophantine approximation, *J. Lond. Math. Soc.* **44** (1991), 1–8.

[82] ———, Hausdorff dimension, lower order and Khintchine's theorem in metric Diophantine approximation, *J. reine angew. Math.* **432** (1992), 69–76.

[83] ———, Geometric and probabilistic ideas in the metrical theory of Diophantine approximation, *Usp. Mat. Nauk* **48** (1993), 77–106, English transl. in *Russian Math. Surveys* **48** (1993), 73–102.

[84] ———, A note on metric inhomogeneous Diophantine approximation, *J. Austral. Math. Soc. (Series A)* **62** (1997), 175–185.

[85] M. M. Dodson and S. Hasan, Systems of linear forms and covers for star bodies, *Acta Arith.* **61** (1992), 119–127.

[86] M. M. Dodson, M. V. Melián, D. Pestana, and S. L. Velani, Patterson measure and ubiquity, *Ann. Acad. Scient. Fenn.* **20** (1995), 37–60.

[87] M. M. Dodson, J. Pöschel, B. P. Rynne, and J. A. G. Vickers, Hausdorff dimension of small divisors for lower dimensional KAM-tori, *Proc. Roy. Soc. Lond. A* **439** (1992), 359–371.

[88] M. M. Dodson, B. P. Rynne, and J. A. G. Vickers, Averaging in multi-frequency systems, *Nonlinearity* **2** (1989), 137–148.

[89] ———, Metric Diophantine approximation and Hausdorff dimension on manifolds, *Math. Proc. Cam. Phil. Soc.* **105** (1989), 547–558.

[90] ———, Diophantine approximation and a lower bound for Hausdorff dimension, *Mathematika* **37** (1990), 59–73.

[91] ———, Diophantine approximation by linear forms on manifolds, *Proc. Indian Acad. Sci. (Math. Sci.)* **100** (1990), 221–229.

[92] ———, Dirichlet's theorem and Diophantine approximation on manifolds, *J. Number Th.* **36** (1990), 85–88.

[93] ———, Khintchine-type theorems on manifolds, *Acta Arith.* **57** (1991), 115–130.

[94] ———, The Hausdorff dimension of exceptional sets associated with normal forms, *J. Lond. Math. Soc.* **49** (1994), 614–624.

[95] ———, Simultaneous Diophantine approximation and asymptotic formulae on manifolds, *J. Number Th.* **58** (1996), 298–316.

[96] M. M. Dodson and J. A. G. Vickers, Exceptional sets in Kolmogorov-Arnol'd-Moser theory, *J. Phys. A* **19** (1986), 349–374.

[97] I. R. Dombrovsky, Simultaneous approximation of real numbers by algebraic numbers of bounded degree, *Dokl. Akad. Nauk BSSR* (1989), no. 3, 205–208.

[98] H. G. Eggleston, Sets of fractional dimension which occur in some problems in number theory, *Proc. Lond. Math. Soc.* (1952), 42–93.

[99] K. Falconer, Classes of sets with large intersection, *Mathematika* **32** (1985), 191–205.

[100] ———, *The geometry of fractal sets*, Cambridge Tracts in Math., vol. 99, Cambridge University Press, 1985.

[101] ———, *Fractal geometry*, John Wiley, 1989.

[102] ———, *Representation of families of sets, dimension spectra and Diophantine approximation*, preprint, University of St Andrews, 1998.

[103] H. Federer, *Geometric measure theory*, Springer-Verlag, 1969.

[104] J.-L. Fernandez and M. V. Melián, Bounded geodesics of Riemann surfaces and hyperbolic manifolds, *Trans. Amer. Math. Soc.* **9** (1995), 3533–3549.

[105] J.-L. Fernandez and D. Pestana, Distortion of boundary sets under inner functions and applications, *Indiana Univ. Math. J.* **41** (1992), 439–448.

[106] L. R. Ford, A geometrical proof of a theorem of Hurwitz, *Proc. Edinburgh Math. Soc.* **35** (1917), 59–65.

[107] D. H. Fowler, Mathematics of Plato's Academy, Clarendon Press, 1987, augmented second edition, 1999.

[108] P. Gallagher, Metric simultaneous Diophantine approximation, *J. Lond. Math. Soc.* **37** (1962), 387–390.

[109] T. Gramchev and M. Yoshino, *Rapidly convergent iteration method for simultaneous normal forms of commuting maps*, preprint, 1998.

[110] P. Gruber, In most cases approximation is irregular, *Rend. Sem. Mat. Univers. Politecn. Torino* **41** (1983), 19–33.

[111] P. Gruber and C. G. Lekkerkerker, *The geometry of numbers*, North-Holland, 1987.

[112] V. Guillemin and A. Pollack, *Differential topology*, Prentice-Hall, 1974.

[113] R. Güting, On Mahler's function θ_1, *Mich. Math. J.* **10** (1963), 161–179.

[114] G. H. Hardy and E. M. Wright, *An introduction to the theory of numbers*, 4th ed., Clarendon Press, 1960.

[115] G. Harman, *Metric number theory*, LMS Monographs New Series, vol. 18, Clarendon Press, 1998.

[116] F. Hausdorff, Dimension und äusseres Mass, *Math. Ann.* **79** (1919), 157–179.

[117] W. K. Hayman and P. B. Kennedy, *Subharmonic functions*, vol. 1, Lond. Math. Soc. Monographs, no. 9, Academic Press, 1976.

[118] M. R. Herman, Recent results and some open questions on Siegel's linearisation theorem on germs of complex analytic diffeomorphisms of $\mathbb{C}^n$ near a fixed point, *Proceedings of the VIIIth International Congress of Mathematical Physics*, Marseilles 1986 (M. Mebkhout and R. Seneor, eds.), World Scientific, 1987, pp. 138–184.

[119] R. Hill and S. L. Velani, *Diophantine approximation in Julia sets of sets of parabolic rational maps*, In preparation.

[120] ———, The shrinking target problem for matrix transformations of tori, *J. Lond. Math. Soc.*, to appear.

[121] ———, Ergodic theory of shrinking targets, *Invent. Math.* **119** (1995), 175–198.

[122] ———, Metric Diophantine approximation in Julia sets of expanding rational maps, *Publ. Math. I.H.E.S.* (1997), no. 85, 193–216.

[123] ———, The Jarník-Besicovitch theorem for geometrically finite Kleinian groups, *Proc. Lond. Math. Soc.* **77** (1998), 524–550.

[124] T. Hinokuma and H. Shiga, Hausdorff dimension of sets arising in Diophantine approximation, *Kodai Math. J.* **19** (1996), 365–377.

[125] V. Jarník, Zur metrischen Theorie der diophantischen Approximationen, *Prace Mat.-Fiz.* (1928–9), 91–106.
[126] ———, Diophantischen Approximationen und Hausdorffsches Mass, *Mat. Sbornik* **36** (1929), 371–382.
[127] ———, Über die simultanen diophantischen Approximationen, *Math. Z.* **33** (1931), 503–543.
[128] ———, Sur une théorème de Mahler, *Casopis* **68** (1939), 59–60, see also V. Jarník, *Remarques à l'article précédent de M. Mahler*, Casopis **68** (1939), 103–111.
[129] ———, Über einen übertragungssatz, *Monatsh. Math.* **48** (1939), 277–287.
[130] F. Kasch, Über metrischen Eigenschaft der S-Zahlen, *Math. Z.* **70** (1958), 263–270.
[131] F. Kasch and B. Volkmann, Zur Mahlerschen Vermutung über S-Zahlen, *Math. Ann.* **136** (1958), 442–453.
[132] R. Kaufman, On the theorem of Jarník and Besicovitch, *Acta Arith.* **39** (1981), 265–267.
[133] J. Kelley, *General topology*, Van Nostrand, 1955.
[134] A. I. Khintchine, Einige Sätze über Kettenbruche, mit Anwendungen auf die Theorie der Diophantischen Approximationen, *Math. Ann.* **92** (1924), 115–125.
[135] ———, Über die angenäherte Auflösung linearer Gleichungen in ganzen Zahlen, *Rec. math. Soc. Moscou Bd.* **32** (1925), 203–218.
[136] ———, Zwei Bermerkungen zu einer Arbeit des Herrn Perron, *Math. Z.* **22** (1925), 274–284.
[137] ———, Zur metrischen Theorie der diophantischen Approximationen, *Math. Z.* **24** (1926), 706–714.
[138] D. Y. Kleinbock, Badly approximable systems of affine forms, *J. Number Th.*, to appear.
[139] D. Y. Kleinbock and G. A. Margulis, *Flows on homogeneous spaces and Diophantine approximation on manifolds*, Ann. Math. **148** (1998), 339–360.
[140] S. Kobayashi and K. Nomizu, *Foundations of differential geometry*, Interscience, 1963.
[141] S. Kochen and C. Stone, A note on the Borel-Cantelli lemma, *Ill. J. Math.* **8** (1964), 248–251.
[142] J. F. Koksma, *Diophantische Approximationen*, Ergebnisse d. Math. u. ihrer Grenzgebiete, vol. 4, Springer, 1936.
[143] E. I. Kovalevskaya, A geometric property of extremal surfaces, *Mat. Zametki* **23** (1978), 99–101.
[144] E. I. Kovalevskaya and N. V. Sakovich, An analogue of Pyartli's theorem for complex analytic functions, *Vestī Akad. Navuk Belarus, Fīz.-Mat.* **4** (1994), 16–20.
[145] J. Kubilius, On an application of I. M. Vinogradov's method to the solution of a problem of the metrical theory of numbers, *Dokl. Akad. Nauk SSSR* **67** (1949), 783–786.
[146] ———, On a metrical theorem in the theory of Diophantine approximation, *Trudy Akad. Nauk Litov. SSR Ser. B* **18** (1959), 3–7.
[147] L. Kuipers and H. Niederreiter, *Uniform distribution of sequences*, Wiley–Interscience, 1974.
[148] J. Lehner, *Discontinuous groups and automorphic functions*, Amer. Math. Soc., 1964.
[149] J. Levesley, A general inhomogeneous Jarník-Besicovitch theorem, *J. Number Th.* **71** (1998), 65–80.
[150] P. Lochak, Hamiltonian perturbation theory, periodic orbits, resonance, intermittency, *Nonlinearity* **6** (1993), 885–904.
[151] E. Lutz, *Sur les approximations diophantiennes linéaires et p-adiques*, Hermann, 1955.
[152] K. Mahler, Über das Mass der Menge aller S-Zahlen, *Math. Ann.* **106** (1932), 131–139.
[153] ———, Über diophantische Approximation in gebiete der p-adic Zahlen, *Jber. Deutsches Math. Verein.* **44** (1934), 250–255.
[154] G. A. Margulis, Formes quadratiques indéfinies et flots unipotents sur l'espaces homogènes, *C. R. Acad. Sci., Paris, Ser. I* **304** (1987), 249–253.

[155] N. I. Markovich, On the approximation of zero by the values of quadratic polynomials, *Izv. Akad. Nauk BSSR, Series Fīz.-Mat.* (1986), no. 4, 18–22.

[156] V. I. Mashanov, Baker's problem in the metric theory of Diophantine approximation, *Izv. Akad. Nauk BSSR, Series Fīz.-Mat.* (1987), no. 1, 34–38.

[157] P. Mattila, *Geometry of sets and measures in Euclidean space*, Cambridge University Press, 1995.

[158] M. V. Melián and D. Pestana, Geodesic excursions into cusps in finite volume hyperbolic manifolds, *Mich. Math. J.* **40** (1993), 77–93.

[159] M. V. Melián and S. L. Velani, Geodesic excursions into cusps in hyperbolic manifolds, *Math. Gött.* (1993), no. 45.

[160] Y. V. Melnichuk, Diophantine approximation on curves and Hausdorff dimension, *Dokl. Akad. Nauk Ukrain. SSR Series A* **9** (1978), 793–796.

[161] ———, Diophantine approximations on a circle and Hausdorff dimension, *Mat. Zametki* **26** (1979), 347–354, English transl. in *Math. Notes* **26** (1980), 666-670.

[162] ———, Hausdorff dimension in Diophantine approximation of p-adic numbers, *Ukrain. Mat. Zh.* **32** (1980), 118–124.

[163] J. W. Milnor, *Topology from a differentiable viewpoint*, The University Press of Virginia, 1965.

[164] I. L. Morotskaya, Hausdorff dimension and Diophantine approximation in $\mathbb{Q}_p$, *Dokl. Akad. Navuk BSSR* **31** (1987), 597–600.

[165] I. M. Morozova, Diophantine approximation by lacunary polynomials with small derivatives at a root, *Papers in honour of V. G. Sprindžuk's 60th birthday,* Inst. Math., Belarus Acad. Sc., 1997, pp. 75–77.

[166] J. Moser, On invariant curves of area-preserving maps of the annulus, *Nachr. Akad. Wiss. Gött., Math. Phys. Kl.* (1962), 1–20.

[167] ———, *Stable and random motions in dynamical systems*, Princeton University Press, 1973.

[168] ———, Is the solar system stable?, *Math. Intelligencer* **1** (1978), 65–71.

[169] ———, On commuting circle mappings and simultaneous Diophantine approximation, *Math. Z.* **205** (1990), 105–121.

[170] J. Nash, The imbedding problem for Riemannian manifolds, *Ann. Math.* **63** (1956), 20–64.

[171] N. N. Nehorošev, An exponential estimate of the time of stability of nearly integrable Hamiltonian systems, *Usp. Mat. Nauk* **32** (1977), 5–66, 287, English transl. in *Russian Math. Surveys* **32** (1977), 1–65.

[172] ———, An exponential estimate of the time of stability of nearly integrable Hamiltonian systems. II, *Trudy Sem. Petrovsk.* (1979), no. 5, 5–50.

[173] A. I. Neishtadt, On averaging in systems with several frequencies II, *Dokl. Akad. Nauk* **226** (1976), 1295–1298.

[174] P. J. Nicholls, *The ergodic theory of discrete groups*, LMS Lecture Notes, vol. 143, Cambridge University Press, 1989.

[175] Z. Nitecki, *Differentiable dynamics*, MIT Press, 1971.

[176] B. Novák, Remark on periodic solutions of a linear wave equation in one dimension, *Comm. Math. Uni. Carolinae* **15** (1974), 513–519.

[177] S. J. Patterson, Diophantine approximation in Fuchsian groups, *Phil. Trans. Roy. Soc. Lond. A* **262** (1976), 527–563.

[178] ———, The limit set of a Fuchsian group, *Acta Math.* **136** (1976), 241–273.

[179] ———, *An introduction to the theory of the Riemann zeta-function*, Cambridge University Press, 1988.

[180] ———, Metric Diophantine approximation of quadratic forms, *Number theory and dynamical systems* (M. M Dodson and J. A. G. Vickers, eds.), LMS Lecture Notes, vol. 134, Cambridge University Press, 1989, pp. 37–48.

[181] K. I. Petersen, *Ergodic theory*, Cambridge University Press, 1963.
[182] H. Poincaré, *Les Méthodes nouvelles de la mécanique céleste*, vol. I, Dover Publications, 1957.
[183] A. D. Pollington and R. C. Vaughan, The k-dimensional Duffin and Schaeffer conjecture, *Mathematika* **37** (1990), 190–200.
[184] Ch. Pommerenke, *Univalent functions*, Vandenhoeck and Ruprecht, 1975.
[185] D. de S. Price, Gears from the Greeks, *Trans. Amer. Phil. Soc.* **64** (1974), 1–70.
[186] B. I. Ptashnik, *Improper boundary problems for partial differential equations*, Naukova Dumka, 1984.
[187] A. S. Pyartli, Diophantine approximations on submanifolds of Euclidean space, *Funkts. Anal. Prilosz.* **3** (1969), 59–62, English translation in *Functional Anal. Appl.* **3** (1970), 303–306.
[188] R. A. Rankin, Diophantine approximation and the horocyclic group, *Can. J. Math.* **9** (1957), 277–290.
[189] C. A. Rogers, *Hausdorff measure*, Cambridge University Press, 1970.
[190] H. R. Rüssmann, On the existence of invariant curves of twist mappings of the annulus, *Geometric dynamics*, Lecture Notes in Mathematics, vol. 1007, Springer-Verlag, 1983, pp. 677–712.
[191] ———, *Invariant tori in the perturbation of weakly non-degenerate integrable Hamiltonian systems*, preprint Fach. Math. Nr. 14, Uni. Mainz, 1998.
[192] B. P. Rynne, The Hausdorff dimension of certain sets arising from Diophantine approximation by restricted sequences of integer vectors, *Acta Arith.* **61** (1992), 69–81.
[193] ———, Regular and ubiquitous systems, and $\mathcal{M}_\infty^s$-dense sequences, *Mathematika* **39** (1992), 234–243.
[194] ———, Hausdorff dimension and generalised simultaneous Diophantine approximation, *Bull. Lond. Math. Soc.* **30** (1998), 365–376.
[195] W. H. Schikhof, *Ultrametric calculus*, Cambridge University Press, 1984.
[196] W. M. Schmidt, A metrical theorem in Diophantine approximation, *Can. J. Math.* **12** (1960), 619–631.
[197] ———, Metrical theorems on fractional parts of sequences, *Trans. Amer. Math. Soc.* **110** (1964), 493–518.
[198] ———, Metrische Sätze über simultane Approximation abhängiger Grössen, *Monatsh. Math.* **68** (1964), 154–166.
[199] ———, Über Gitterpunkte auf gewissen Flächen, *Monatsh. Math.* **68** (1964), 59–74.
[200] ———, On badly approximable systems of linear forms, *J. Number Th.* **1** (1969), 139–154.
[201] ———, *Diophantine approximation*, Lecture Notes in Mathematics, vol. 785, Springer-Verlag, 1980.
[202] C. Series, The modular surface and continued fractions, *J. Lond. Math. Soc.* **31** (1985), 69–80.
[203] C. L. Siegel, Iteration of analytic functions, *Ann. Math.* **43** (1942), 607–612.
[204] ———, Über die Normalform analytischer Differentialgleichungen in der Nähe einer Gleichgewichtslösung, *Nachr. Akad. Wiss. Gött. Math-Phys. Kl* (1952), 21–30.
[205] D. C. Spencer, The lattice points of tetrahedra, *J. Math. Phys.* **21** (1942), 189–197.
[206] M. Spivak, *A comprehensive introduction to differential geometry*, vol. 3, Publish or Perish, 1979.
[207] V. G. Sprindžuk, A proof of Mahler's conjecture on the measure of the set of S-numbers, *Amer. Math. Soc. Transl. Ser.* 2 **51** (1966), 215–272.
[208] ———, *Mahler's problem in metric number theory*, Translations of mathematical monographs, vol. 25, American Mathematical Society, 1969, Translated by B. Volkmann.

[209] ———, New applications of analytic and p-adic methods in Diophantine approximations, *Actes Congrès Inter. Math.* **1** (1970), 157–160.

[210] ———, *Metric theory of Diophantine approximations*, John Wiley, 1979, Translated by R. A. Silverman.

[211] ———, Achievements and problems in Diophantine approximation theory, *Usp. Mat. Nauk* **35** (1980), 3–68, English transl. in *Russian Math. Surveys*, **35** (1980), 1–80.

[212] B. Stratmann and S. L. Velani, The Patterson measure for geometrically finite groups with parabolic elements, new and old, *Proc. Lond. Math. Soc.* **71** (1995), 197–220.

[213] D. Sullivan, Conformal dynamical systems, *Proceedings of the Conference on Geometrical Dynamics, Rio de Janeiro*, Lecture Notes in Mathematics, vol. 1007, Springer-Verlag, 1981, pp. 725–752.

[214] ———, Disjoint spheres, approximation by imaginary numbers, and the logarithm law for geodesics, *Acta Math.* **149** (1982), 215–237.

[215] ———, Entropy, Hausdorff measures old and new, and the limit set of geometrically finite Kleinian groups, *Acta Math.* **153** (1984), 259–277.

[216] R. Taylor and A. Wiles, Ring-theoretic properties of certain Hecke algebras, *Ann. Math.* **141** (1995), 553–572.

[217] E. C. Titchmarsh, *The theory of the Riemann zeta function*, Oxford University Press, 1951.

[218] P. Tukia, The Hausdorff dimension of the limit set of a geometrically finite Kleinian groups, *Acta Math.* **152** (1984), 127–140.

[219] ———, On the dimension of limit sets of geometrically finite Möbius groups, *Ann. Acad. Sci. Fenn. Ser. A I Math.* **19** (1994), 11–24.

[220] M. Urbánski, The Hausdorff dimension of the set of points with non-dense orbit under a hyperbolic dynamical system, *Nonlinearity* **4** (1991), 385–397.

[221] D. V. Vasilyev, Diophantine sets in $\mathbb{C}$ and Hausdorff dimension, *Papers in honour of V. G. Sprindžuk's 60th birthday*, Inst. Math., Belarus Acad. Sc., 1997, pp. 21–28.

[222] S. L. Velani, Diophantine approximation and Hausdorff dimension in Fuchsian groups, *Math. Proc. Cam. Phil. Soc.* **113** (1993), 343–354.

[223] ———, An application of metric Diophantine approximation in hyperbolic space to quadratic forms, *Publ. Math.* **38** (1994), 175–185.

[224] ———, Geometrically finite groups, Khintchine-type theorems and Hausdorff dimension, *Math. Proc. Cam. Phil. Soc.* **120** (1996), 647–662.

[225] J. A. G. Vickers, Infinite dimensional inverse function theorems and small divisors, *Number theory and dynamical systems* (M. M. Dodson and J. A. V. Vickers, eds.), Cambridge University Press, 1989, pp. 19–36.

[226] A. I. Vinogradov and G. V. Chudnovsky, The proof of extremality of certain manifolds, *Contributions to the theory of transcendental numbers* (G. V. Chudnovsky, ed.), Amer. Math. Soc., 1984, pp. 421–447.

[227] I. M. Vinogradov, *Elements of number theory*, Dover Publications, 1954.

[228] B. Volkmann, Zur Mahlerschen Vermutung im Komplexen, *Math. Ann.* **140** (1960), 351–359.

[229] H. Wegmann, Das Hausdorff-Mass von Cantormengen, *Math. Ann.* **193** (1971), 7–20.

[230] A. Wiles, Modular elliptic curves and Fermat's last theorem, *Ann. Math.* **141** (1995), 443–551.

[231] T. Willmore, *Differential geometry*, Oxford University Press, 1959.

[232] E. Wirsing, Approximation mit algebraischen Zahlen beschränkten Grades, *J. reine angew. Math.* **206** (1961), 67–77.

[233] J. C. Yoccoz, *An introduction to small divisors problems*, From number theory to physics, Springer-Verlag, 1992, Les Houches, 1989, pp. 659–679.

[234] ———, *Petits diviseurs en dimension 1*, Astérisque **231** (1995).

[235] Kunrui Yu, A note on a problem of Baker in metrical number theory, *Math. Proc. Cam. Phil. Soc.* **90** (1981), 215–227.

[236] F. F. Zheludevich, Simultane diophantische Approximationen abhängiger Grössen in mehreren Metriken, *Acta Arith.* **45** (1986), 87–98.

Index